"WE SHALL FIGHT IN THE STREETS"

by **Captain S. J. Cuthbert,**
Scots Guards

A GUIDE TO STREET FIGHTING—GROUND—DEFENCE—ATTACK—USE OF EXPLOSIVES—ARMS AND EQUIPMENT—TRAINING—EXERCISES.

"To say that street fighting has been neglected is an understatement too dangerous to be pleasant" says the author in his Preface. "The possession of a town maintains a grip upon the lifeline that can strangle an offensive. Towns are the ready-made answer to the Blitzkreig. Properly held they must slow an advance."

Many valuable lessons were learned during the late war and street fighting is still something which must be considered in training as of primary importance. Street fighting is carried out in unique and unnatural conditions and it is necessary to analyse and understand the peculiar features which go to make street fighting such a highly skilled form of warfare.

Here is a book which is a valuable addition to the Library of any Unit and for the Regimental Instructor who is faced with the problem of teaching the rudiments of street fighting.

"WE SHALL FIGHT
IN THE STREETS!"

"We shall fight on the seas and oceans, we shall fight with growing confidence and growing strength in the air, we shall defend our Island whatever the cost may be. We shall fight on the beaches, we shall fight on the landing-grounds, we shall fight in the fields and in the streets, we shall fight in the hills; we shall never surrender."

"We shall defend every village, every town, and every city. The vast mass of London itself, fought street by street, could easily devour an entire hostile army."

"WE SHALL FIGHT IN THE STREETS"

Guide to

STREET FIGHTING

GROUND — DEFENCE — ATTACK — USE OF
EXPLOSIVES—ARMS AND EQUIPMENT—
TRAINING—EXERCISES

BY

CAPTAIN S. J. CUTHBERT
Scots Guards
ILLUSTRATED BY JOHN G. WALTER

The Naval & Military Press Ltd

Published by

The Naval & Military Press Ltd
Unit 5 Riverside, Brambleside
Bellbrook Industrial Estate
Uckfield, East Sussex
TN22 1QQ England

Tel: +44 (0)1825 749494

www.naval-military-press.com
www.nmarchive.com

In reprinting in facsimile from the original, any imperfections are inevitably reproduced and the quality may fall short of modern type and cartographic standards.

CONTENTS

CHAPTERS

	PAGE
CHAPTER 1: THE GROUND	1
CHAPTER 2: THE DEFENCE	3
CHAPTER 3: THE ATTACK	10
CHAPTER 4: THE USE OF EXPLOSIVE	21
CHAPTER 5: ARMS AND EQUIPMENT	25
CHAPTER 6: TRAINING	28
CHAPTER 7: EXERCISES	33

APPENDICES

	PAGE
APPENDIX 1: PROTECTION TABLE	45
APPENDIX 2: THE STRENGTHENING AND ADAPTATION OF A HOUSE FOR DEFENCE	46
APPENDIX 3: SYLLABUS FOR A DEMONSTRATION	51
APPENDIX 4: SUPPRESSION OF CIVIL DISTURBANCES	57

PREFACE

To say that street fighting has been neglected is an understatement too dangerous to be pleasant.

Consider, on the one hand, the military importance of towns. Leave out of account the fact that all governments reside in towns, that all vital industries are concentrated in towns, that all military dumps and depots lie in towns. Forget that nearly all good landing-places round our shores are, because they *are* good landing-places, built over; or that the Navy depends for its command of the oceans upon ports and naval bases, one and all of them cities. Overlooking these, remember this one fact, namely, that all road and rail communications pass through towns. Stop, and digest this fact. Think for a moment of the system of supply to an army; then return to this fact —all road and rail communications pass through towns.

Consider, on the other hand, the advantages of the defence in towns. Notice how in the recent war towns have held out long after the country round them has been overrun. Consider how the enemy official doctrine teaches that towns shall be by-passed and attacked later by troops specially detached for the purpose. They recognize that a town is a strong point, and they follow their doctrine of by-passing the strong point. But note that they also recognize the necessity of attacking the town. The reason is this: that both their armoured divisions and their infantry divisions depend for supplies upon wheeled vehicles: those vehicles run on roads and rails; and —all road and railway communications pass through towns.

No doubt towns are vulnerable to air and artillery bombardment. Experience seems, however, to point to the need for great air superiority if a town is to be made untenable. For examples of this, we need not look to Madrid, Tobruk or Odessa; we can remember the daylight *Blitz* on London. No one who witnessed Cockney courage and stoicism is likely to forget it; but the decisive fact which saved London was the German failure to gain air supremacy. And if an overwhelming concentration of air and artillery force has to be collected each time a town is encountered, lightning advances will be a thing of the past.

Possession of a town maintains a grip upon the lifeline that can strangle an offensive. Towns are the ready-made answer to the *Blitzkrieg*. Properly held they *must* slow an advance.

This book has been written for the Regular Army and for civilians. It is not the result of actual fighting experience: it is the result only of experience in training. It is certain, therefore, that some of the ideas and methods suggested are mistaken. Please read it with an alert and challenging mind. If I am wrong, it is you who will die, not I.

Chapter IV has been written by Captain N. MacBean, R.E., late of Eastern and South-Eastern Command Weapon Training School, Dorking.

I would like to say how grateful I am to Mr. and Miss Benfield, of Streatham, for their great help and encouragement in the early stages of this book.

I must express my thanks to Mr. C. W. E. Remnant, F.S.I., for his kind and much-needed information on house construction. Also the officers of the Small Arms School, Bisley, for their invaluable advice upon arms and equipment.

CHAPTER I

THE GROUND

CONDITIONS dictate methods: the fittest survive, because they best understand and adapt themselves to the conditions in which they live.

Street fighting is carried on in unique, unnatural conditions; *only* the fittest survive. It is above all necessary to analyse and understand the peculiar features of ground which go to make street fighting such a highly skilled form of warfare.

It is difficult in thinking of cities to discard the many familiar details and to focus upon the simple features which alone are of military significance.

   i. The ground upon which most towns are built was the ground that covers England; the ordinary slightly rolling land, intersected by rivers, streams and hedgerows, and patched with woods. The woods and hedges have for the most part been cut to make way for masses of human habitations, but every town retains small areas of park and woodland.
  ii. Vast human effort has gone to the perfection of communications in urban areas. Today towns may, generally speaking, be said to consist of, alternately, rows of houses and means of communication to them, here and there cut by railways, canals and other lines of communication.
 iii. In the masses of homes that have been built, wealth has drawn distinctions.

The rich, though they may work in towns, generally choose to live in the country. Where, owing to the size of a city, they cannot easily escape, they occupy a central position. They build their houses strong and big, either detached and standing in their own grounds, or in rows surrounding a railed square of lawns and gardens.

The houses of the poor are usually situated near a great factory or industrial centre. Normally they consist of lines of small, two- or three-storeyed houses, ill-built, on either side of wide, straight streets, backed by little yards and the rear of another row of houses. These yards are usually a conglomeration of fences, sheds, shelters, bins, and other forms of obstacle and cover from view.

From the early nineteenth century onwards, unprecedented growth of population and industry demanded masses of hastily built houses for the new middle and lower middle classes. The advent of town planning and the fact that whole areas were built over by one firm gave us the modern suburbs which have grown round the outside of nearly every town. They consist of rows of small "semi-detached" or detached, lightly built, two-storeyed houses, standing between front and back gardens, fitting in a large design of criss-crossing streets. An enormous number of hedges and fences have been built to enclose these properties, so that back gardens offer both good cover from view and, cumulatively, great obstacles to an advance.

Spaced throughout the cities are shopping and commercial centres, usually consisting of very tall, strongly built houses, flanking main arterial streets. In many cities building has followed the haphazard lay-out of old centres of commerce, and we find the same buildings facing each other across narrow, winding streets, flanked by alleyways and passages.

Whatever the class of house or area, three generalizations can be made which are of the most vital importance, giving rise as they do to the basic principles of street warfare. It is these three features which must be understood, and digested:

i. No other type of country is either so open or so close. In every street are coverless stretches, ideal fields of fire, death-traps to the unwary attackers. Bordering every street are a hundred protected firing positions, a hundred hiding-places, a hundred ambush positions.

ii. It is possible to climb 30, 50, perhaps 100 feet in as many seconds. Street fighting thus possesses a third dimension, not often present in field warfare.

iii. Cities present exceptionally blind and disjointing conditions. In no other form of warfare are there such narrow horizons, or such ruthless divisions between units of the same force.

In this chapter it is not intended to discuss the rules of good street fighting. One such may, however, be given pride of place, for upon its unvarying observance depends not merely the life of any particular soldier but the attainment of the object he is ordered to achieve.

It has been said above "No other country is either so open or so close." It is well-nigh impossible to be more than five yards from cover.

This is the rule, and it is shared with field warfare; but it is more urgent, more emphatic!

USE COVER. NEVER HANG ABOUT IN THE OPEN!

CHAPTER II

THE DEFENCE

1. General

In the first chapter, the military features of built-up areas were analysed. In this and the following chapters conclusions are drawn from them, which form, so to speak, rules of conduct: and suggested methods of obedience to them are laid down. How many of these methods are put into practice must depend upon the precise nature of the attack and the time allowed for defence.

2. Forms of Attack

In this country two conditions of attack seem likely to present themselves. For the first, the regular land advance, the defender will probably have ample time to prepare a formidable reception.

The second type of attack allows no time for preparation. A company of enemy parachutists can land on a green space three hundred yards square and be ready to move or fight in under fifteen minutes. This will allow no time for elaborate booby traps, barricades or wiring.

3. Distribution of Forces

The first problem facing any unit commander is that of the distribution of his forces. The terrain offers the attacker a great variety of covered approaches, all of which must normally be stopped. If, however, he merely divides his force equally between them he will succeed only in weakening the defence of them all. It follows that he must weigh the needs of a fixed defence against those of a mobile reserve.

The object of having a mobile reserve is, of course, to be able to concentrate a force at a point attacked by the enemy, either for counter-attack or reinforcement of the local defenders.

Against a land attack there may be time to make elaborate preparations for defence. If there is such time the defence can be made overwhelmingly strong and a large mobile reserve should not be necessary.

Against an attack by paratroops it is most unlikely that any preparations for a fixed defence could be made. The method forced on the defenders would seem to be to contain them by lines of "stops" and then clear the area enclosed—in other words, to become the attackers.

It should always be borne in mind that except at night or in thick fog, or where an ideal covered line of approach exists which the enemy are unlikely to cover, the sacrifice of the advantage of stillness conferred upon the defence is justified only by necessity, or to obtain surprise against a badly trained enemy.

4. Siting the Defence

In placing the members of any unit, the following points should be borne in mind.

  i. The system of defence must pivot round the L.M.G. In built-up areas fields of fire tend to be very narrow. The high rate of fire of the L.M.G. makes it a weapon of inestimable value.

  ii. The 360° field round any house is normally split into several narrow fields of fire. Riflemen should be used to cover all but the most important approaches.

  iii. "Interior lines" abound in built-up areas. The distance between the defence of one street and the defence of the next is often the width of a house. Full advantage should be taken of this.

  iv. It is easier to fire to one's left than to one's right. When firing to the left, the rifle comes out of a window before the body: when firing to the right, the body must be exposed before the rifle can be brought to bear.

  v. In every street there are hiding-places in which men may lie low until the enemy has passed, and open fire from behind. If these men wish to fire to their left they must be on the side opposite to those prepared to fire on an approaching enemy. These two parties can therefore have the additional task of covering each other's doorways.

  vi. Two men are far more than twice as strong as one, both for psychological reasons and because there are four sides to a house and at least two lines of approach.

  On the other hand, a whole section placed in one house will find themselves surrounded without having been able to fire, unless sections are very plentiful.

  It can be taken as a guide, therefore, that a house should be held by never less than two men and rarely more than five.

  vii. The L.M.G. should be sited in a house selected for its dominant field of fire, its strength of construction and its unobtrusive position. Normally the house which is selected is at the end of a street, looking down the street, satisfying the first condition at the expense of the third.

  From a corner house it is possible to fire in three directions. Although often desirable, such a selection does not avoid the obvious: the position must therefore be made very strong, and alternative positions prepared.

  A house lying farther back from the road than its neighbours will often have the priceless advantage of being shielded by them from enemy covering fire.

  viii. German street-fighting tactics are based on an attack from the rear. It is essential, above all else, to have all-round defence.

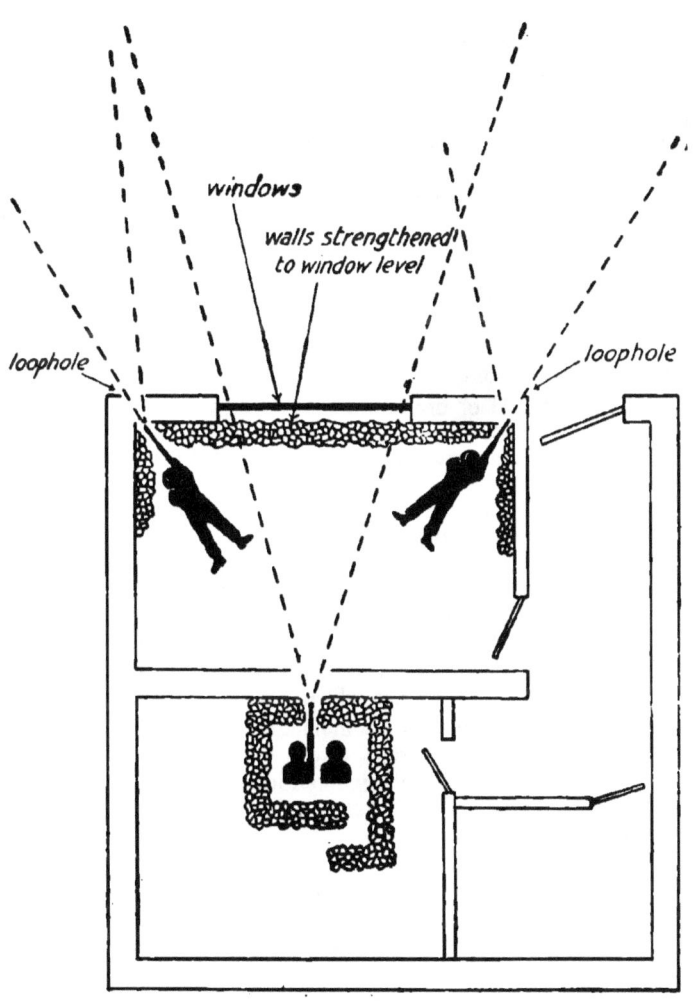

Method of widening field of fire, while ensuring unobtrusiveness.
N.B. Barricades, mines, & other personnel are not shown.

5. Preparation of a House for Defence

Given time, there is practically no limit to the number of improvements which can be made to a defensive position in a house. Below is a long list of possible improvements. It is obvious that in very many cases time will not allow for all of them to be carried out. It should be equally obvious that it would be criminal on that account to ignore them.

Fire positions have been selected:
  i. For their good field of fire.
  ii. For strength.
  iii. For unobtrusiveness.

These three characteristics of a good defensive position should be exaggerated in every conceivable way.

i. GOOD FIELD OF FIRE.
  (a) The best field of fire is obtained from the lowest possible firing position. Many houses have basements whose windows just clear the surface of the ground: these should be used if possible.
  In other cases a ventilator may be enlarged or a loophole knocked in the wall.
  (b) The farther forward the L.M.G. is placed the wider will be the field of fire: on the other hand, the more vulnerable will it be. If it is necessary to have a wider field of fire, loopholes may be made in the side walls for riflemen.

ii. STRENGTH.—Every fire position can, given time, be made proof against the collapse of the house in which it is sited.
  (a) Remove breakables and inflammables, *e.g.*, window glass, ornaments, curtains, rugs, etc.
  (b) Send a man round the neighbourhood commandeering fire extinguishers, water buckets, etc.
  (c) Shore up the fire position.
  (d) Build an emplacement.
    1. These can be made of sandbags or rubble between boards, in chests of drawers, etc. The emplacement should be roofed over with timber: joists taken from another room are suitable.
    2. The emplacement must have a solid foundation. It will generally be necessary to remove part of the floor (see Appendix II).
    3. A man fires over the following heights:
    Rifle or L.M.G.:

Standing	4 ft. 6 in.
Kneeling or sitting	30 in.
Lying	12 in.

M.G.:
    Sitting   ..   ..   24 in.

4. Inside dimensions:
    6 ft. long.
    3 ft. 6 in. wide.
    5 ft. high.

5. Loopholes: build them with narrow exits, widening towards the inside to allow for traverse. Duplicate them freely, as they are impossible to hide.

(e) All entrances must be barricaded or, where it is impossible to hide a barricade, mined. This applies to all doors and windows. Booby traps should be laid in the hall and on the staircase.

(f) A line of withdrawal must be left. In the case of attached houses, go next door, find a cupboard on the party wall, and knock a hole in the back into the defended house; then close the cupboard. With detached houses knock a hole into a lean-to shed or bush: or lean a wheelbarrow, bin or other article against the hole. If necessary place a dummy booby trap over the hole.

(g) Knock loopholes in walls to cover every approach to the house: a loophole from the occupied room to cover the front hall is also useful.

(h) Find some fine wire-mesh and wire over all the windows to prevent grenades being thrown in. Cut a slit in the wire to allow the dropping out of grenades.

(i) Make a dummy position. A bolster hung on a wire at the back of a room three doors away can be connected by a wire run through the intervening walls. A jerk may deceive, and will certainy distract, enemy observers.

(j) Prepare an alternative position. The Germans used tracked guns to blow a house to pieces; no amount of strengthening will withstand a direct hit.

(k) Make a gas-proof room and stock it with candles (the electric light is bound to fail), torches, water, medical and food supplies, and ammunition.

iii. UNOBTRUSIVENESS.—When all improvements to the field of fire and strengthening of the house are complete, it is essential to remove any clues which may betray your position to the enemy.

There are two principal methods:
  (a) *Concealment*.
    Be sure there is no external sign of a barricade.

Remove the trail of sand and rubble which almost certainly leads up to the door.

Put muslin curtains over the windows. They are opaque from the outside and transparent from the inside—besides being the very symbol of respectability.

(b) *Duplication.*

Work which cannot be hidden must be duplicated.

If yours is the only door in a damaged street which is tight shut (to conceal a barricade), shut others in the street.

Clear of glass the windows in half a dozen houses, and wire them over.

Make dummy loopholes in your own house and in the houses whose doors you have closed and windows you have cleared of glass.

If the field of fire has had to be cleared so much as to become obvious, treat the other houses in the same way.

6. Positions other than in Houses

Having considered the defence of a house at such length, one is apt to forget that there are in built-up areas other excellent positions. Slit trenches offer better cover against air and artillery bombardment than any house: they are not, however, useful against a land attack, as they can be commanded from the upper floors of neighbouring houses.

In street fighting one expects the enemy to take refuge in houses. A thick hedge, the branches of a tree, a heap of refuse, a pile of rubble may all provide excellent surprise positions. It must be quite clear, however, that most of these positions depend upon concealment: discovery will render one as vulnerable as if one were standing motionless in the middle of a street.

7. Anti-Tank Defences

Tanks suffer from serious handicaps in built-up areas, for the following reasons:

i. The enemy can erect road blocks which cannot be seen from a distance.

ii. The enemy can remain concealed until a tank is directly beneath his position.

iii. Tanks cannot fire more than, at an average, 30 degrees above the horizontal.

iv. Visual contact between tank units is lost, and direction is very hard to keep.

It is clear, however, that they suffer from these handicaps *only* as long as the enemy is prepared to take advantage of them.

The most suitable positions for an ambush are:
  i. On the upper floors of a strongly built house.
  ii. Above a road block or a point where the enemy is likely to stop at first sight of a road block.

The same weapons may be used against tanks in streets as in the fields. The third dimension in street fighting and the fixed nature of the defence are ideal for the use of the heavier grenades, such as the No. 73 or the S.I.P. grenade.

For a list of anti-tank weapons and their characteristics see M.T.P. No. 42 (1940).

8. Street Obstacles

  i. Trenches, or barricades, of wagons, cars, furniture, etc., may be very useful in holding up enemy troops or vehicles, especially if placed just round a corner to effect surprise. They are not, however, effective as fire positions because they can be commanded from the tops of buildings. They should be covered by fire from a neighbouring building and not used as a breastwork.
  ii. Wire is the best possible obstacle to advancing troops. A street properly wired and covered by an L.M.G. is a death-trap. It forces the attacker to commit the great, and invariably the last, sin of street fighting—hanging about in the open. Wire should be placed far enough away from the defended house to prevent the enemy throwing grenades and should not be so laid as to give away the position of the house to the enemy. Tins with pebbles inside should be hung on the wire to raise the alarm should the enemy try to remove it under cover of darkness.

CHAPTER III

THE ATTACK

1. Forms of Attack

Broadly, attacks in built-up areas have one or both of two main objects:
   i. To penetrate to a certain area.
   ii. To clear a certain area of enemy.

It is impossible to say what precise forms attacks in cities or towns will or should take. They will vary with the object, the ground, and other particular circumstances.

The following is a general impression of the lines which a modern attack may follow, and is intended more to stimulate thought than to prophesy events.

The process may be one of neutralization and infiltration.
   i. The town is subjected to heavy bombardment by aircraft or artillery, or is sprayed with gas. The main enemy body may by-pass the town, leaving its capture to special troops detailed for the purpose. The latter will probably attack the town as soon as the bombardment ceases, from any direction they judge to be favourable.
   ii. As a result of reconnaissance and careful study of the town plan, certain few buildings or localities well inside the town may be selected as first objectives. At first small and then large parties of troops will force their way to these objectives to form strong points from which the core of the town may be eaten out and any perimeter defences attacked from the rear. The objectives will be selected so as to be co-operative, *i.e.*, success in training objective A will aid those troops who are trying to gain objective B.

2. Methods of Attack

The scheme of attack tabulated below has been laid down *not* in order to dictate hard-and-fast rules to an attacking force but in order that a clear idea may be obtained of the problems facing an attacker and of a reasonable solution to them. It will almost always be impossible for the attacker to keep completely to the ideal. The chaos attending street fighting, and the haste imposed upon all commanders will demand a compromise: if this compromise is to be successful, initiative and a high degree of training are required of all ranks.

Failure to realize the inevitability of disorganization and chaos would be dangerous. It must be accepted that commanders will lose control of their units, that members of the same unit will lose touch with each other and may even fire upon one another, that direction will be lost and pockets of enemy exist behind the forward attacking troops.

Under such circumstances:

i. A simple plan is essential. Each unit must be given a clear limited objective, and complicated manœuvres such as a change of direction should generally be avoided. Isolated thrusts apart from the main attack are not usually successful and will not divert as many defenders from the main attack as might be achieved in field warfare, since the bulk of the defence will probably be on a fixed system giving all-round defence.

ii. Success or failure will largely depend upon the initiative of subordinate commanders. Lacking clear orders as to their next actions, section and platoon commanders must on no acount merely send back each time to be told what to do, but must act swiftly and with enterprise in the spirit of their previous orders. Speed in attack must not be allowed to flag for an instant. The need for a clear limit of exploitation, however, referred to in sub-para. (i) will now be even more clearly seen.

iii. Accurate, early information is vital. A plan made with insufficient information will fail; a plan made with incorrect information will fail. A sub-unit commander has not done his duty if his superior commander does not know everything about the enemy that he knows, and all that is required about his position, intention, casualties and state of exhaustion.

iv. Tremendous thrust is required. In this difficult terrain, the attackers are forced to move in the open against defenders who, given time to choose and perfect their fire positions, can make them feel like rabbits in a wood, continually surprised and shot at from unexpected directions. The attackers can upset the scales only by giving their opponents no time to prepare, no rest in retreat, no chance to reorganize or reassure themselves. Speed is of the essence, and speed can be produced only by intelligent anticipation, efficient orders, dashing execution, and energy, more energy, and energy again.

3. Advance

It is usually impossible absolutely to safeguard the main body against attack. An active advance guard will, however, use some of the many good points of observation that abound in cities, cover with fire the more important junctions, and question inhabitants.

The main body should be well deployed. It may often be best to keep all sections on one side of the road, for two reasons:
  i. They can keep to the shadow.
  ii. In the event of a sudden attack, they will all take cover on the same side of the street, and will not therefore be divided from each other by a coverless, bullet-swept area.

4. Protection

From the start to the end of operations there must be no moment during which the main body can be surprised undeployed. On arrival at the assembly position, whether on foot or in M.T., O.Ps. and sentries will be posted covering every approach. Two A.A. sentries per company should be posted in an open space or square covering opposite arcs of 180°. The main body should take immediate cover and be slightly dispersed.

5. Reconnaissance

Horizons of brick and mortar make reconnaissance difficult. The ideal of an aeroplane view being usually denied, second best is a view from a high building. Other useful aids are street maps, aeroplane photographs, and the accounts of troops on the spot, and local inhabitants. Officers commanding troops likely to be used in built-up areas may find it useful to obtain aeroplane photographs of their area for use instead of maps.

6. Plan

In forming his plan the commander must bear the following points in mind:
  i. A complicated plan depending for its success upon co-ordination between troops attacking up widely separated streets, or upon changes of direction, is liable to fail.
  ii. Streets provide easier and quicker lines of approach and attack. Back gardens may sometimes offer more cover; but it should be remembered that it will often be necessary to climb over obstacles, and at these moments exposure is unavoidable.
  iii. A single attack should be maintained. This does not mean that only one street or one house is to be attacked at a time. It does mean that if two or three or six streets are attacked at a time the effort should be co-operative, *i.e.*, success in street 5 helps the troops attacking street 1.

7. Headquarters

Headquarters should be chosen for the following features:

  i. Well up: much closer to the forward troops than in field warfare.
  ii. Easy to find.
  iii. A strongly built house or shelter.
  iv. Entrance and exit defiladed from enemy view and fire.

No concentration of troops or M.T. must be allowed near it, by day or by night. During an attack it may frequently be necessary for the Headquarters to be moved. Chalk or other marks may be drawn as a guide to the new Headquarters. For safety it can be arranged beforehand that, *e.g.*, arrows face the opposite to the true direction.

8. Cordons

The object of cordoning is to prevent enemy lateral movement, whether of reinforcement, counter-attack or escape.

When the object of an attack is to clear an area of enemy, the procedure should be to cordon a sub-area, clear it, and keep it cordoned from the uncleared area.

When the object of an attack is to penetrate to an area, it may frequently be advisable to cordon the line of advance.

Cordoning should be carried out by fire, not by men. The firing position should be selected so as to give the best field of fire, and will therefore normally be as low as possible.

9. Covering Fire

It should be an absolute rule that no troops move across the open until three measures have been taken:

  i. A rifle or L.M.G. must be sited to cover the street in which they are moving. In choosing the firing position, bear in mind:

   (*a*) That if the enemy see you getting into position or if your position is the obvious one to choose, you will never fire from it.

   (*b*) That S.A. fire penetrates up to $13\frac{1}{2}$ inches and .55-inch anti-tank rifle fire up to 27 inches of brickwork, and that prolonged bursts will achieve greater penetration.

   (*c*) That the lower you are the longer will be the beaten zone of your weapon.

   (*d*) That if you are going to cover the advance of your own troops from the rear you must get high enough up to fire over their heads until the last possible moment.

(e) That there are many good firing positions other than in houses.
   (f) That fire from a position rather above a room occupied by the enemy will search the position best and is the most telling.

   For details of fire positions from a house, see Chapter II.
ii. One L.M.G. per company should be sited on one of the highest buildings in the area to cover as far as possible all rooftops. It is very hard for troops in streets to deal themselves with attackers above them. No street should be attacked without domination of the rooftops.
iii. There is usually a house whose windows dominate a whole street. It may be placed at the end of the street, or on rising ground farther away. If there is such a house, make preparations before attacking the street to destroy or blind it if the enemy should open fire from it.

10. The Point Section

The first troops to break cover are the point section. This consists of:

i. *Scouts.*—The duty of the scouts is to find out where the enemy is. They move, one or two on each side of the street, by short bounds, stopping under the cover of doorways, buttresses, gateposts, alleyways, etc., to observe the opposite side of the street. When moving, they should keep as close to the side wall as possible, and must go at the highest possible speed. During their pauses for observation they must on no account keep their heads poking round a corner; if possible they should be equipped with periscopes. If fired on, the scouts should take cover and try to work into position to give covering fire to the assault party.

ii. *Observers.*—Behind the scouts move the observers. Their duty is to observe and report on any enemy movements. The normal number is two for each side of the road. This is a minimum, for should there be anything to report one must go back, leaving one to observe. Like the scouts, they should move rapidly, and close to the side walls. They should be so far up that they can see what is happening to the scouts; and so far back that they are not exposed to the same fire as the scouts, and can get back to report. If they are pinned to cover by enemy fire they can usually shout back a message. A catapult is a possible method of sending back a message to the end of the street.

In the case of the enemy opening fire the message should contain:

(a) Strength and armament of enemy.

(b) Exact position, side of street, number or description of house, first or second floor, right or left window.

(c) Whether a scout is in position to give covering fire.

(d) What possible covered approach there is to the house.

iii. *Remainder of the Section.*—Under the N.C.O. the remainder of the point section works along behind the scouts and observers, helping to search out the enemy or protect the section. A man may be sent down a cul-de-sac or a little passageway between houses to look to a flank. Another may be posted to cover a street running off to a flank. A third may be used to cover the scouts and observers when they have got rather far from their previous covering fire.

It may be necessary to strengthen the point section. This should be done when the point section is likely to have far to go.

11. Mopping Up

It must not be thought that because the point section has encountered no opposition a street is clear of enemy. In every street there are enough covered positions to conceal a battalion.

If the street is to be cleared, as soon as the point section has covered a sufficient distance:

i. Arrangements should be made to cover the backs of the houses on each side of the street with fire, preferably L.M.G. fire.

ii. The two rear sections should advance, one on each side of the street, searching the houses. The N.C.O. should remain in the street controlling operations, while his men, working in pairs, pass rapidly from house to house. The search need not be thorough, for a single enemy concealed in a cupboard will be cut off from his friends and in no mood for a stout resistance.

12. Consolidation

It is not usually enough to clear a street; it must be held, to prevent enemy reoccupation or counter-attack. Normally only a single rifleman can be spared, for there may be many streets, and men and weapons will be at a premium.

13. Assault

i. *Forms of Attack.*—When the enemy is encountered he must be dealt with. There are many forms of attack. He may be

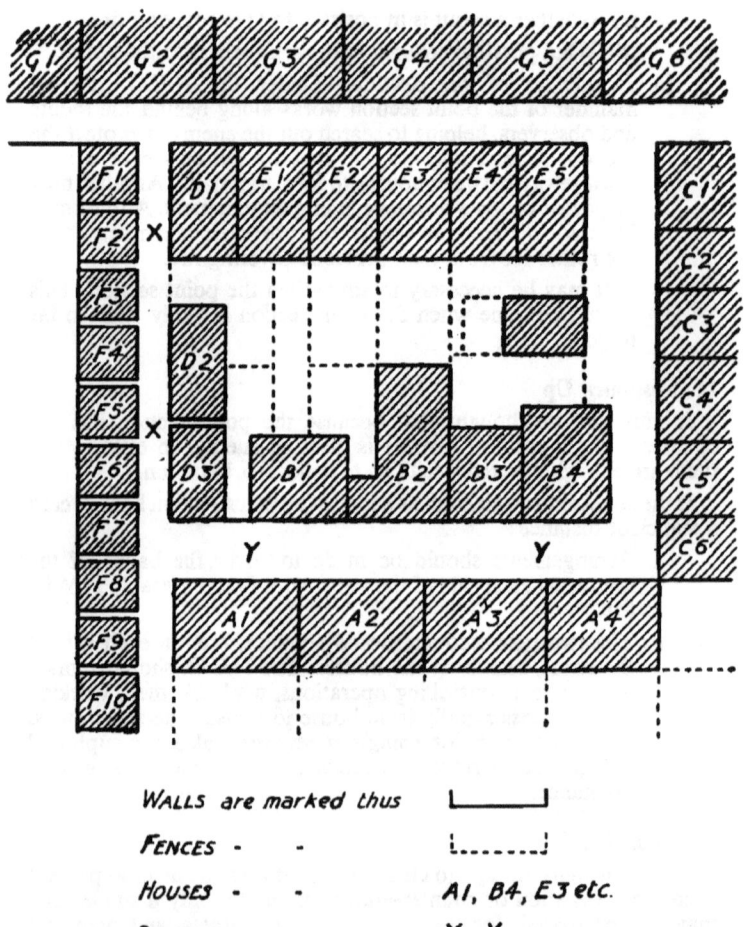

	WALLS are marked thus	⌊___⌋
	FENCES - -	⌊......⌋
	HOUSES - -	A1, B4, E3 etc.
	STREETS - " - ...	X, Y.

attacked with bombs from the air; though this form is not practicable in close fighting, and is more useful against an area than against an individual target. He may be attacked with shells from a gun man-handled into position; with A.P. weapons capable of riddling his protective cover; with incendiary weapons fired or thrown; with gas; with *any* effective weapon from a gun to a hosepipe. Often, however, it will not be possible to oust or destroy the enemy without an infantry assault, and it is the infantry assault, whether combined with other forms of attack or not, with which we are concerned.

ii. *The Lines of Assault.*—In the same way, there are many lines of assault. An assault may be made up main streets; up side streets, through back gardens; over rooftops; through neighbouring houses; through a house opposite and across a road; possibly even through sewers or contiguous cellars. Gliders were used by the Germans in Belgium to land on a wide, flat roof at Fort Eban Emaal. We may leave out of account these more exotic forms of attack while remembering that if they are possible to do they will be a surprise, a *new idea,* and therefore successful. In deciding upon a line or lines of assault the following points should be borne in mind:

(*a*) An enemy position is a strong point. Before it is attacked it should be as nearly encircled as possible. Occupy the houses on either side or behind it.

(*b*) Gain the houses on one side of a street before you enter the street. Example (see diagram opposite): A and F are occupied by our own troops. Enemy in B will find it hard to cover the B side of the street and will have to lean right out of the windows, doors, etc., to do so, where they will be themselves exposed from A2, 3 and 4. Thus the street can only be effectively covered by the enemy from C, and this narrows the opposition almost to a single house, which can be neutralized.

(*c*) Attack the enemy most threatened by your own position. Example (see diagram opposite): A and F are held by own troops. Which street should be made good first— X or Y?

Analysis: one side of each street is held. One end and one side of each street is held by the enemy. Of the enemy houses holding X, D2 and D3 are covered by fire on four sides, and D1 is covered on two sides. Of the enemy houses holding Y, four sides of all the houses D3 and B1, 2, 3 and 4 are covered by fire. Moreover, the street joining X from the enemy side is not properly covered, whereas the entrances to Y are well covered. Conclusion: street Y should be attacked first.

(d) More than one line of attack will be found difficult to co-ordinate but will have an excellent psychological effect upon the enemy.

(e) If you can attack along a line which does not hinder your covering fire the latter will be more effective.

(f) Back gardens may offer good cover: advancing troops will, however, find it difficult to avoid exposing themselves when surmounting obstacles.

(g) It is hard to deal with attackers from above; rooftops are safe from below, and, being built as a rule in the shape of an inverted V, they have usually a side defiladed from fire.

(h) In the case of an attack through houses, back doors a little distance from the enemy house are usually covered from view by high wooden fencing, sheds, etc. Moreover, in "jerry-built" houses it does not take more than a minute or two to knock a hole in the wall.

(i) The most suicidal occupation in war is to delay in the open.

(j) The enemy are unlikely to occupy a single house unsupported by any other position. Normally they may occupy houses on each side of the road and site their L.M.G. in a house whose windows dominate the whole street (see Chapter II).

iii. *The Assault.*—Experience in Madrid showed that a very high proportion of the total casualties was incurred in assault, and of these 75 per cent. were caused by delay outside the barricaded house of the enemy. The whole assault must be carried through at the highest possible speed by the shortest exposed route, with the heaviest possible covering fire until the last possible moment.

The following points should be considered:

(a) The enemy will prefer firing to his left. Firing off the right shoulder he has to expose his body less when firing to the left than when firing to the right. Incidentally, the greater the angle to his position, both lateral and vertical, at which he has to fire, the more must he expose his body.

(b) The enemy will undoubtedly have barricaded or mined all entrances to his house. Preparations must be made to force an entrance to the house with the shortest delay outside. The best method is for a single man to run forward under covering fire, place an explosive against the house and dodge under cover. He should avoid doors and windows in placing his explosive; possibly a point in the side wall is best. (For the most suitable explosives and methods of use, see Chapter IV.)

If explosives are not obtainable, second-best are a heavy axe and a crowbar. Their employment means delay in the open and should be avoided whenever possible. Then men who are using them should always be accompanied by a man watching for bombs and grenades thrown from windows or a roof, ready to seize and throw them over a wall or down the street.

If explosives are used to force an entrance, the remainder of the assault party must be under cover well up, ready to dash in before the enemy has recovered from the shock of the explosion. Whatever the method employed of forcing an entrance, the assault party must attempt all methods of getting in. They need not necessarily confine themselves to going in through the gap they have made; the enemy will be concentrating their attention on it, and a man going in over the roof of a lean-to shed, up a drainpipe or down the skylight may escape notice.

(c) *Clearing the House.*—Having made an entrance, the assault party still has half its task before it. Before leaving cover to assault the house, the officer or N.C.O. in charge must allot tasks to each member of his party to ensure that there is no hanging about inside the house and that no room, cellar or attic is unsearched.

The following points should be borne in mind:

(1) The house should be searched systematically, floor by floor, remembering the cellars and roof. It is better if possible to search a house from the top downwards, as this makes an ally of gravity for grenades, etc. It does, however, involve entrance from the roof.

(2) The front hall and the staircase are the two most dangerous points. Try any method of getting up and down from floor to floor but the main staircase. If it is not barricaded it will certainly be mined. The fire escape may help. If the staircase must be used, feel in front of you with a long pole or throw a grenade at it, to set off any booby traps.

(3) The officer or N.C.O. must stay in the hall or passage to direct operations, but he must make use of any cover there may be, *e.g.*, a thick sofa.

(4) If the enemy are known to be upstairs, a few bursts of S.A. fire through the floor from underneath will be a prudent action.

(5) Smoke grenades inside a house are quite overwhelming. Incendiary bottles or bombs are very intimidating. (For arms and equipment suitable, see Chapter V).

(6) Prisoners can be used to clear the house by setting off or revealing booby traps.

(7) Enter a room crouching as low as possible. The enemy will be prepared to fire about three or four feet above the floor.

It is essential that all attackers are familiar with methods of defence outlined in the chapters on defence and booby traps.

CHAPTER IV

THE USE OF EXPLOSIVE

1. General

The use of high explosives can solve many of the problems that arise during street fighting with an ease that astounds those who have never seen this weapon in action. It is essential, however, that only those who are trained in the preparation and application of H.E. charges are permitted to handle it.

The training itself is a simple matter and requires only average intelligence and determination on the part of the student. Methodical habits combined with reasonable enthusiasm are preferable to headstrong and suicidal dash, as, both in preparation and handling, explosives call for care and a steady head lest they become a source of danger to the user and his comrades. Arrangements should be made to have two or three men from each platoon trained in this subject, as they will be invaluable when the opportunity to use H.E. arises.

In the course of this chapter high explosives only are referred to, a classification which excludes gunpowder. Methods of handling are not given, as practical training is essential in this subject. The chapter confines itself to suggesting some of the tactical uses of H.E., both in defence and in attack.

2. Types of Explosive

The principal explosives now in use are:

i. *Gelignite.*—A plastic explosive of high power, suitable for all cutting operations, usually coloured buff, grey, brown or yellow. It is commonly supplied in 4-oz. sticks and can be initiated by a detonator only. It requires careful storage to avoid deterioration, and has definite temperature limits of 32° F. and 110° F. It must always be stored in a dry atmosphere. When placed in position and left for any length of time it must be well protected from damp.

ii. *Guncotton.*—The safest of all explosives in Service use, it requires a primer to initiate it. Packed in air-tight tins it can be stored more easily than gelignite and it does not deteriorate readily. It is not so powerful as gelignite, but can be used for all cutting operations, the main disadvantage being that it is issued in 1-lb. blocks and is not plastic.

iii. *T.N.T.*—A compressed powder issued in $1\frac{1}{4}$-lb. blocks, it is comparable in use and application with guncotton. One block of T.N.T. will have equal effect with one block of guncotton. T.N.T. is easy to store, but has the disadvantage that the blocks are brittle and crumble easily. It requires a primer to initiate.

iv. *Explosive 808.*—An extremely powerful explosive issued in 4-oz. sticks. It has better storage qualities than gelignite and is even more powerful, but it requires a primer to initiate it.

v. *Ammonal.*—A grey powder, mainly used for cratering and other mining operations, it is the least powerful of all Service explosives. Its slow speed makes it most suitable for shifting large masses of earth, but it should never be used for a cutting or shattering operation. Ammonal must be kept perfectly dry, as the least degree of damp renders it useless. It is issued in 25-lb. and 50-lb. tins and also in 4-oz. waterproof cartridges.

3. Defence

As explained in Chapter II, it is unlikely that enemy tank formations will venture into built-up areas; it is more probable that infantry will be called upon to capture a town.

All houses which are adapted for defence should be fitted with a booby-trap system, operated electrically. The whole system can be controlled by a master switch which is situated so that the last defender can make the system live as he leaves. Booby traps will not be suspected in a house which has been used for defence, and the ensuing casualties will cause the attackers to reduce their speed of advance—even if it does not force them out into the open. Alternatively, each house can be mined either by buried charges or by explosives concealed in the basement or ground floor. These charges can be fired from the neighbouring houses, and a single house blown up in this way will deter attackers from entering any other house near by with any feeling of confidence.

For details of booby traps see Field Engineering Pamphlet No. 9, Plates 16, 17, 18, 19, 23, 24 and 25, and the written material referring to these plates.

4. Explosives in Attack

When clearing the enemy from occupied houses the advantages of explosives are very evident. The approach to an occupied house can be made on a blind side, *i.e.*, the side where there are no downstairs windows. The house can be entered by a hole blown in this blind side, the charge required in most cases being surprisingly small.

If an entrance is made immediately after the explosion it will be found that the defenders in the immediate vicinity of the breach will be either casualties or in no condition to put up any resistance. Once inside the house, similar methods can be adopted to break into any room that offers resistance. Further, if a charge of 1 lb. of H.E. on a 5-second or 7-second fuse be thrown into a room occupied by the enemy, the mopping-up operation becomes extremely simple and safe.

German troops were trained in the use of pole charges, *i.e.*, charges of 10 lb. or more of H.E. on the end of a long pole, fired by an ordinary time fuse. These pole charges were intended to be placed in the loopholes of pill-boxes and were carried forward under cover of smoke or darkness. An adaptation of these charges can be made with advantage, and men should be trained to apply such charges to street fighting when clearing points strongly held by the enemy.

With an elementary knowledge of explosives it is a comparatively simple matter to clear buildings which can block a field of fire.

5. Anti-Tank Measures

Though the unsuitability of tanks in built-up areas has been stressed, it must on no account be taken that tanks will not be used by the enemy. Tanks are only unsuitable as long as every precaution has been taken against them. Every preparation should be made for their reception, and in this explosives play a large part.

Charges should be buried across all approach roads at a depth of 9 inches or a foot. The quantity of this should be 4 lb. per foot of length. The charge must be fired electrically and provision can be made to fire this from the cover of a neighbouring house. This charge will wreck lorries, but will only smash the track of a tank and bring it to a standstill. Though badly shaken, the crew will still be in a condition to fight and their guns will still be in action. To finish off the tank it is necessary to get a charge of explosive on the hull of the tank itself.

This can be done under cover of smoke or the charge can be lowered from the upper floors of houses near by. Arrangements can be made to have charges suspended above the road at suitable points or they may be lowered with the aid of long poles. The charges should, if possible, be placed on the rear deck of the tank behind the turret: though the thickness of the armour plating in this position varies, a charge of 10 lb. should smash a hole in the hull and disable or kill the crew. It is desirable to have some knowledge of the weakest parts of a tank's plating, and this knowledge should enable the charges to be placed to the best advantage. As an alternative to the continuous road charge, separate charges of 4 lb. can be laid in the same manner as anti-tank mines at close spacing, though they must be buried sufficiently deeply to prevent their being damaged by traffic. The separate charges should be connected up by instantaneous fuse so that they can all be fired simultaneously by a low-powered battery exploder.

In addition to these road-block charges, a number of 10-foot planks should be prepared with the appropriate quantity of explosive strapped to them. These portable road charges can be rushed to any threatened point and laid by the roadside suitably concealed. On the approach of tanks or armoured fighting vehicles these planks can be dragged across the road and exploded under the vehicle as required.

6. Demolition

Small bridges can be destroyed with ease and certainty, and craters blown in the road to form tank barriers. These and many other simple demolitions can be carried out after a short but careful course of training. It must be repeated, however, that H.E. must be handled only by trained men and that men should be trained without further delay. The possibilities of this weapon in trained hands are unlimited, but in the hands of untrained personnel its use is usually disastrous.

To sum up generally, a knowledge of the use of explosives places in the hands of a single individual the most powerful weapon in existence, a weapon which, handled with reasonable care and skill, is practically safe and adaptable to a very high degree. In using explosives it is not necessary to hold large stocks and kinds and sizes of grenades, each grenade being suitable for only one job. A single store of H.E. can be held and by varying the quantity used this one item of store can be adapted to a variety of purposes.

The final, and in fact ruling, point in the handling of H.E. is that personnel must be trained carefully and must at all times continue to observe the precautions taught in this training. Accidents will only create a loss of confidence in the weapon, whereas the loss of confidence should be in the operator. Provided this is borne in mind, the unit commander will find a solution to the majority of street-fighting problems in the use of H.E.

CHAPTER V

ARMS AND EQUIPMENT

General

The choice of arms and equipment should be governed by certain features of street warfare:

1. Street fighting imposes great physical strain upon all ranks.
2. Many obstacles are encountered in built-up areas.
3. Nearly all surfaces are hard and smooth: water drains off quickly.
4. Fighting usually takes place at close quarters.
5. There are usually many firing positions giving good cover from fire: most of these have overhead cover which can quickly be strengthened.

The above points suggest definite conclusions:

1. Arms and equipment should be kept as light as possible.
2. Weapons should correspond to one of the following types:
    i. Giving a great volume of fire in proportion to weight and size, *e.g.*, a sub-machine gun.
    ii. Being able either to get round or to penetrate cover from S.A. fire, *e.g.*, a grenade or Boys rifle.

1. Arms

i. *Rifle and Bayonet* (S.A.T. Nos. 3 and 12.)—The 360° field round any defensive position is likely to be split, in built-up areas, into several separate, narrow fields of fire. The rifle is the ideal weapon to cover the less important fields of fire. The bayonet is the principal close-quarter weapon for the ordinary rifleman, and can be used with great moral and actual effect in clearing houses, or in street *mêlées*.

ii. *Discharger.*—The present issue of one discharger per platoon is intended for the use of the No. 68 grenade (see below). For this purpose one rifle per platoon should be strengthened by binding.

iii. *M.M.G. and L.M.G.* (S.A.T. Nos. 4 and 7).—These two weapons will be dealt with together, as in street fighting the greater range and endurance of the M.M.G. will rarely be utilized.

Both types are excellent in defence and can make a street impassable. The L.M.G. is suitable in attack to give covering fire, but both are too heavy and cumbersome for close-range fighting, such as would be the case in gardens and houses. At a pinch an L.M.G. can be fired from the hip, and may be useful where sub-machine guns are not obtainable.

iv. *The Sub-Machine Gun* (S.A.T. No. 21).—This is the ideal assault and close-fighting weapon. It enables targets appearing from different directions and at short ranges to be engaged quickly, and has a high rate of fire combined with great stopping-power. The sub-machine gun can be fired from hip or shoulder with accuracy up to 50 yards. It has been found that the best method of firing from the shoulder is to align the foresight in the V formed by the cocking-handle rather than the backsight.

v. *The Boys Anti-Tank Rifle* (S.A.T. No. 5).—This has good penetrative qualities (see Appendix I), and, although primarily for use against A.F.Vs., can usefully be used against enemy behind cover proof against S.A. fire. The hole blown by the bullet will not be large enough to provide entry to a house. The rifle is rather heavy and awkward to carry great distances.

vi. *Two-inch Mortar* (S.A.T. No. 8).—This weapon is difficult to use and rather ineffective in built-up areas, for the following reasons:

(*a*) Mortar bombs have little penetration.

(*b*) The enemy will, in defence, be roofed over; and, in attack, will either be roofed over or better engaged by S.A. fire.

There are, however, occasions when it is useful to attack an enemy in gardens, ruined houses, etc., while it may sometimes be possible to use smoke, in cases where an O.P. can be found close to the weapon.

vii. *Pistol.*—In the hands of an untrained shot, the pistol will not prove an effective weapon. With constant practice, it will be found that the short barrel and the speed with which single rounds can be fired enable the firer to engage targets rapidly in different directions and render it a very useful weapon for close-quarter fighting.

viii. *No. 36 Grenade* (S.A.T. No. 13).—This grenade is no longer fired from a discharger, but when thrown is a most valuable weapon. It will normally be used in clearing a house and in surprise encounters with enemy at close quarters.

It has a 4-second fuse and can be thrown up to about 35 yards. The thrower must take cover before the explosion.

No. 36 grenades are *not* suitable as explosives to blow a gap in barricades, doorways, walls, etc.

ix. *No. 68 Grenade.*—This grenade is fired from a discharger on an E.Y. rifle. It has great penetrative power, but will not blow a large hole. The blast effect inside a small room may be very considerable.

x. *No. 69 Grenade.*—This grenade is in a light bakelite container and has a blast effect over a small area. Though its destructive qualities are not great, it has a tremendous moral effect, which makes it a useful weapon for fighting at close quarters. The thrower need not take cover after throwing.

xi. *No. 73 Grenade.*—This grenade, designed for use against A.F.Vs., has great destructive qualities, and will therefore be useful against walls, etc. It can be thrown up to 20 yards. The thrower must take cover.

xii. *The S.I.P. Grenade.*—Incendiarism is an effective form of attack on emplacements, houses and A.F.Vs. The S.I.P. grenade has therefore a distinct value in street fighting. A minor advantage is the smoke it produces, which may temporarily blind the enemy and force him to put on a respirator.

xiii. *Smoke Grenades, Generators, etc.*—The smoke grenade is obsolescent. Smoke generators are at present issued only for training purposes.

Smoke will almost certainly play a most important part. In defence it may be used to blind the enemy covering fire and mystify the attacking troops, incidentally making them good silhouette targets. In attack it may similarly be used to blind enemy fire, but better surprise is often achieved if a side is attacked different from that which has been blinded. In the assault, smoke inside a house can be overwhelming.

2. Equipment

i. *Shoes.*—When obtainable, rubber-soled shoes are in every way preferable to Army boots, combining the advantages of silence and a good grip.

ii. *Respirators and Gas-Capes.*—Respirators may be needed against S.I.P. grenades, dust, etc., and should always be carried. If orders permit, gas-capes may be discarded in built-up areas, where there is so much cover from spray attack.

iii. *Haversack.*—There are, in street fighting, features which make the wearing of the haversack by advancing or attacking troops not only undesirable but unnecessary. The contents of the pack are:

(a) Knife, fork, spoon and mess-tin.

(b) Spare pair of socks.

(c) Groundsheet.

(d) Food.

Shelter is provided by houses, where sufficient utensils and underclothes can also be commandeered. Though food may also be found, some must certainly be taken and may be carried in a pouch or pocket.

iv. *Periscopes.*—Periscopes are invaluable for officers and N.C.Os. and for the point section of each attacking platoon. The present issue should be supplemented by home-made productions.

v. *Ladders.*—Short ladders are often useful in the attack, and one per company may be carried. They can often be commandeered on the spot.

CHAPTER VI

TRAINING

General

Training should follow the sequence:
  Lecture.
  Demonstration.
  Sand table.
  Practice.

1. Lecture

i. A lecture is made more vivid if accompanied by blackboard diagrams or a street sand table (see below).

ii. Tabulate the lecture and help the audience to take notes by showing what your headings are.

iii. Watch your audience to allow them time to take down notes.

iv. Do not read your lecture. You must be vivid, animated, and get your lecture "across."

v. Above all, keep *still* while you are talking. It is impossible to follow a shuttlecock for an hour.

2. Demonstration

N.B.—The following paragraph may frighten some readers off demonstrations for good and all. Precisely the opposite effect is hoped for. There is no form of training so vivid or effective as a well-prepared demonstration. While it is idle to pretend that a good demonstration will not take time and trouble, there are many refinements mentioned here, *e.g.*, the Tannoy loud-speaker, which can be dispensed with for all normal occasions. (An excellent street-fighting demonstration can be given after six hours' preparation.)

The demonstration should follow the lecture as soon as possible.

i. PREPARATION OF THE DEMONSTRATION.

  (*a*) Select the date and time and the performers. Make sure that they are clear for the hours of rehearsal. Have about four spare men attending the whole preparation, to fill the places of men who go sick, etc.

  (*b*) Work out *on paper* a detailed syllabus of the demonstration. This must be related to the ground, to the men who are performing, and to the audience.

Every confusion or gap in the thought of the instructor will be multiplied by the number of men taking part and lead to untold delay and avoidable inefficiency.

(c) Walk the performers round the area, and describe to them generally what is to happen and where each section of the demonstration is to take place.

(d) On the blackboard, or with toy houses, make a plan of the area and go through with your men the sequence of the demonstration. As you come to each stage, make the platoon sergeant detail off the men who are affected, and make them stand up. Explain in detail what is wanted, and tell them the points which it is desired shall be brought out. Answer questions.

(e) When everybody is quite clear in his mind, and not before, take the men out on the ground and rehearse them through the demonstration itself.

(f) When preparing the demonstration, always consider the position of the audience. The best demonstration in the world is valueless if the audience cannot see it properly. The problem becomes acute in a demonstration of the preparation of a house for defence.

(g) With more than one instructor it is possible to divide the demonstration up into sections with a separate instructor, a separate sub-area, and a separate sub-unit. If this is done, much time can be saved, but it is quite essential to rehearse the whole demonstration through at the end until it falls together.

(h) Above all, it is vital that there shall be one O.C. demonstration, whose word is law.

(i) For expert advice as to shoring, clearing floors, building construction, etc., the O.C. demonstration should ring up the local Home Guard, who are sure to have a carpenter and master builder willing to help. Failing that, the local surveyor will be able to suggest somebody.

ii. DETAIL OF THE PLAN OF THE DEMONSTRATION.

A specimen detail will be found in Appendix III.

iii. NECESSARY ARRANGEMENTS.

(a) *Site.*—The best site is a bombed area comprising at least one street 100 yards long, of derelict but safe houses. The road should be a cul-de-sac or one little used by traffic. For information and permission apply to the District Surveyor or Debris Officer (County Council). A bombed area has the advantage that every house can be entered and damaged with impunity. Other possible sites are a large factory, school, or other private area devoid at one time of day of inhabitants. The area must be cordoned to keep off children, who are capable of disrupting an entire demonstration.

(b) *Personnel.*—The minimum number required for a demonstration is:

    1 Instructor.
    6 Enemy.
    24 Own troops.

(c) *Representation of Fire.*

    L.M.Gs.—Blank, or pebbles rattled in tins.

    Rifle.—Blank, or rattling of bolts.

    Explosives.—Large thunderflashes.

    Grenades.—Small thunderflashes.

    Booby traps.—Thunderflashes electrically wired and used in conjunction with a $4\frac{1}{2}$-volt battery. Use a thread which, when touched, sets off an ordinary mouse-trap to make the electrical connection.

All forms of thunderflash can be indented for or bought from Messrs. C. & T. Brock, Crystal Palace Fireworks, Hemel Hempstead (Boxmoor 580).

(d) *Preparation of a Defensive Position.*—For materials, see Appendix III. If a bombed site is being used, nearly all the materials can be borrowed from the repair gang on the spot.

iv. SHOWMANSHIP.

(a) A Tannoy loud-speaker ensures that a large audience can hear. These can be borrowed from an R.A. regiment. If the audience is to move much, the police have Tannoy loud-speakers mounted on cars.

(b) If the audience is to see the inside of a house prepared for defence it is essential to arrange for one-way movement through the house. Since only a small number of the audience can be inside at any one time, the various points to note should be clearly labelled.

(c) Try to keep a light touch and find an amusing climax.

    Example 1.—Defenders produce strong fire on advancing enemy from room just above audience. At crescendo of fire, drop bits of tree and leaf and a dead crow in the road in front of audience.

The setting of Booby Traps on training

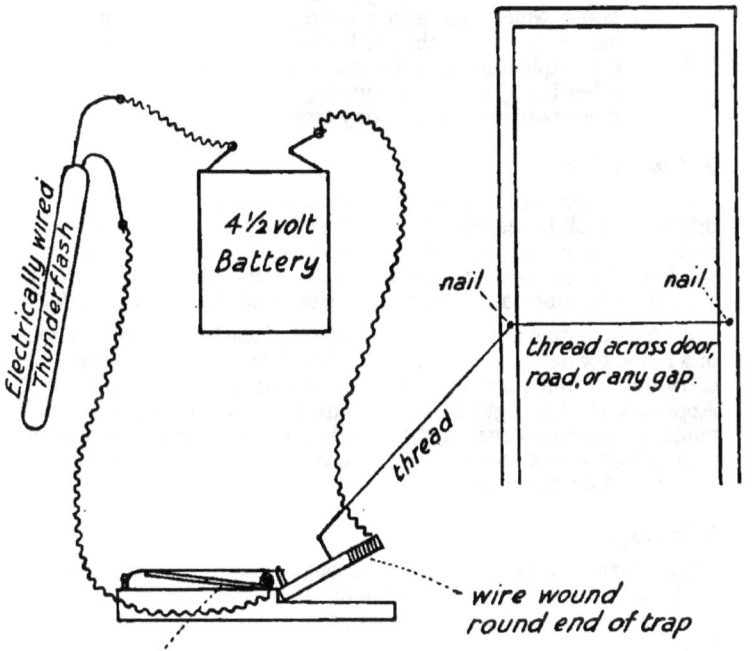

Electrically wired Thunderflash

4½ volt Battery

nail

nail

thread across door, road, or any gap.

thread

wire wound round end of trap

Mouse trap forms contact which is made by trap closing on wire on the wired "cheese" half.

Example 2.—X troops lie low, allow Y troops to go past, then counter-attack and kill them all. X L.M.G., who is giving covering fire, joins in but loses his head and shoots his own troops. All X and Y troops are now draped over the pavement, dead. X L.M.G. No. 1, seeing what he has done, is seized with remorse and shoots himself. Everybody is now dead. Finis.

(*d*) Arrange booby traps to catch the audience on its way home. A thunderflash connected to a steel helmet and respirator, apparently left behind by somebody; to a pickaxe near the house which has been prepared for defence; to a thread across the road; a thunderflash connected to the engine of a car, exploding when the starter is pressed; these and many other booby traps are amusing, easy to lay, and reveal the great possibilities of booby traps.

3. Sand Table

i. Small houses can be bought at most toy shops or made without difficulty. Each house should be at least 3 inches high, 6 inches long and 4 inches broad. The essential point is that entrance, exit and windows should show. About twenty-five houses are sufficient for a sand-table scheme on the usual lines, similar to that in Chapter VII.

ii. A fascinating and instructive method of teaching the adaptation of a house for defence is to buy or make a large-size doll's house and carry out, in miniature, the procedure described in Chapter II and Appendix II. The sight of a miniature room shored up, a row of miniature sandbags against the outside walls, or a miniature booby trap, which works, across a miniature door, is enough to set any students off on their own.

4. Practice

Only first-hand experience can thoroughly imprint on men's minds the lessons taught by lecture, demonstration and sand stable. Continual practice must be given by schemes and T.Es.W.T. for officers and N.C.Os.

For the setting and direction of schemes and T.Eş.W.T. see M.T.P. No. 37, Chapter VII.

CHAPTER VII

EXERCISES

[N.B.—*It is realized that several of the articles of both arms and equipment included in the answers will not normally be obtainable. They are included so that a clear idea may be received of their employment, should they become available.*]

GENERAL NARRATIVE

(See plan at end of book.)

BRITISH forces invading occupied France have reached the town of LANGRES through which runs the arterial road DIEPPE—PARIS. The main forces push on: 1 Guards Brigade is detached and given the special role of capturing the town and opening up the road.

The Brigade Commander divides the operation into three phases:

1. The gaining of two objectives, the SCHOOL and the FIRE STATION by 1 SCOTS GUARDS supported by 1 WARWICKS.
2. The passing up to these objectives of the remainder of the Brigade.
3. The clearance of the rest of the town from these strong points.

NARRATIVE 1

1 S.G. attacks and penetrates as far as street F, having met with slight opposition. There the forward company is held up at nightfall: at 2300 hrs. O.C. 1 S.G. summons his "O" Group.

Extracts from orders of O.C. 1 S.G. in which you are O.C. No. 1 Company. (*You are to assume that O.C. 1 S.G. does not necessarily know very much about street fighting.*)

Information

Enemy.—Enemy appear not to be numerous. No positions have been located.

Own Tps.—1 WARWICKS are supporting us. 1 COLDSTREAM in reserve.

Intention

1 SCOTS GUARDS will capture and occupy the SCHOOL and FIRE STATION.

Method

Battalion will attack along two lines:
1. *Objective.*—SCHOOL.
   No. 1 Coy. forward. No. 2 Coy. in reserve.
   Route.—Street A.
   Boundaries.—Rt. excl. Street B.
   Left excl. Street C.
   Action on Capture of Objective.—No. 1 Coy., consolidation in SCHOOL area.
   Support.—Under command each company:
   One Det. 3-inch Mortars.
   One Sec. Carrier Platoon.
   Coy. Assembly Position.—Street Z.
   Start Line.—Street F.
   Zero hour.—0525 hrs.
2. (Not shown. FIRE STATION is 250 yards due west of SCHOOL. Nos. 3 and 4 Coys.' attack is synchronized with attack of Nos. 1 and 2 Coys. on SCHOOL.)

Administration

Food.—Coys. will be fed by 0430 hrs.
Tpt.—"A" Echelon will be sent to Coy. Assembly Areas by 0400 hrs. Will R.V. at "B" Echelon under T.O. by 0500 hrs.
"B" Echelon remains in wood ½ mile south of ESTAMINET ANGLAIS. Cookers will be sent to Coy. Assembly Position by 0345 hrs.
Ammunition.—(Not shown).
Weapons.—(Not shown).
Equipment.—(Not shown).
R.A.P.—Hospital at 123456.

Intercommunication

Battalion H.Q.—Hospital 123456.
Success Signal.
   No. 1 Coy.—Red, Red, Red.
   No. 3 Coy.—Green, Green, Green.
Synchronize watches.
Any questions?

Question 1.—As O.C. No. 1 Coy. give your orders for the attack. (Fill up the administration paragraph as you think it should be.)

Work out your own solution first, then turn to page 36 and work on the D.S. solution.

Question 2.—As O.C. No. 8 Pl. give your orders for the first phase of the attack only.

After you have worked out your solution, use the D.S. solution on page 38.

NARRATIVE 2

The attack duly starts at 0525 hrs. Almost at once one of your scouts is shot and killed from the upper window at X (see diagram).

Question 3.—What message would you expect to receive back?

Question 3a.—Who would bring it?

See D.S. solution on page 40.

Question 4.—What action do you take?

See page 41 for D.S. solution.

NARRATIVE 3

The first objectives were captured. The forward platoons are attacking in the second phase.

Question 5.—As N.C.O. i/c No. 7 Sec., No. 9 Pl., you have been ordered to assault the SCHOOL, force an entrance in the wall opposite House No. 4 and make good one room into which Nos. 8 and 9 Secs. may enter. You have been given one No. 73 grenade which, it is calculated, will blow a hole 3 feet square in the wall; and six No. 36 grenades.

What orders have you given your Section at the Assembly Post? (Exclude orders for Phase 1.)

For D.S. solution see page 41.

NARRATIVE 4

The operation is successful, and the second phase is completed. The SCHOOL and the FIRE STATION are strong points. The clearance of the town is to begin. O.C. No. 2 Coy., 1 WARWICKS, is detailed to clear the area bounded by Streets C, F. and B, which has not been held.

Question 6.—Describe how, as O.C. No. 2 Coy., 1 WARWICKS, you would set about clearing the area allotted to you.

(You are to assume:
  i. That the SCOTS GUARDS are unable to help by cordoning or covering fire.
  ii. That your Coy. is at present in the SCHOOL.)

For D.S. solution see page 42.

D.S. SOLUTION TO QUESTION 1

[QUESTION 1.—*As O.C. No. 1 Coy. give your orders for the attack.*]

O.C. No. 1 COY.'S ORDERS

Information

Enemy.—You have seen what opposition we have had so far. Nothing further is known of the enemy.

Own Tps.[1]—Nos. 3 and 4 Coys. attack the FIRE STATION 250 yards west of the SCHOOL at the same time as we attack. No. 2 Coy. is in reserve behind us. We have under command one Det. Mortars, one Sec. Carrier Pl.

Intention

No. 1 Coy. will capture and occupy SCHOOL.

Method

Coy. will attack in two phases.

Phase 1

Coy. will attack two Pls. forward, one in reserve. No. 7 Right, No. 8 Left, No. 9 Reserve.

Objectives.

No. 7 Pl.—Houses Nos. 34, 31, 21, 20, 15. *Route:* back gardens between Streets A and B.

No. 8 Pl.—Houses Nos. 8, 9, 10 and 11. *Route:* Street A.

No. 9 Pl.—Reserve in House No. 39.

Phase 2

Coy. will attack two Pls. forward, one reserve. No. 7 Right, No. 9 Left, No. 8 Reserve.

Objectives.

No. 7 Pl.—House No. 4 and houses beyond SCHOOL. *Route:* Street A.

No. 9 Pl.—SCHOOL. Entrance to be made between House No. 4 and School. *Route:* Street G between Houses Nos. 14 and 8, and 7 and 8.

Leading troops of No. 9 Pl. will not cross Street D until No. 7 Pl. has occupied House No. 4.

---

[1] The Company must know that Nos. 3 and 4 Companies are attacking at the same time, on their left. If they did not know, they would be apt to shoot any troops they met on their flank.

No. 8 Pl. (Reserve).—Houses forming first objective, giving covering fire. Will give smoke to help Nos. 7 and 9 Pls. across Street D.²

*Coy. Boundaries.*³—Right excl. Street B.
　　　　　　　　　　Left. excl. Street C.

Support.
*Carrier Section.*⁴—Carriers in reserve at Coy. Ass. Pos. Sec. will mount one Bren gun on roof of House No. 42 (tallest house) to cover rooftops, and one each cordoning Streets B and C.
*3-inch Mortar Detachment.*⁵—In reserve at Coy. Ass. Pos. Will arrange local defence for Carrier and Mortar Dets.

F.U.P.—Houses in rear of Start Line 1.

Start Lines.—1, Street F.
　　　　　　　　2, Street D.

*Action on Capture of Objective.*⁶
No. 7 Pl.—Hold northern approaches to SCHOOL.
No. 8 Pl.—Hold southern approaches to SCHOOL.
No. 9 Pl.—Hold SCHOOL.

Zero Hour.—0525 hrs.

---

² Troops are not likely to be able to cross a lateral street such as this without either a prolonged smoke screen or neutralization of enemy fire. A well-placed L.M.G. can pass a line of fire down a street as lethal and continuous as a high-tension cable.

³ Boundaries must be given to companies. This is not always practicable with platoons. It is a general rule that troops may cross their boundaries to deal with fire that is holding them up only if:
　1. They cannot get on by using another route within the boundaries allotted to them.
　2. There are no other troops operating on the flank from which the fire comes.

⁴ Although carriers were used with success in France as a surprise street-fighting weapon, they would seem to be most unsuitable against an enemy who has had time to prepare any ambushes or road blocks. If the C.O. makes the mistake of allotting carriers out to companies, Company Commanders will no doubt yield to the temptation to use their weapons dismounted, in some sort of cordoning or covering fire employment. They could also be used for the local defence of a H.Q. area.

⁵ 3-inch Mortar bombs may be useful in a very open area, such as a large goods yard, or from a hill overlooking a built-up area; but normally observation is difficult, while their penetrative qualities are small. The C.O. should have kept them in Battalion reserve.

⁶ Action on capture of objective must be clearly laid down. If the actual houses to be occupied are not specified by the Company Commander, Platoon Commanders must liaise, to avoid confusion.

Administration

Food.—Coy. feed at Assembly Position at 0350 hrs.[7]

Tpt.

"*A*" *Echelon.*—Haversacks, boots, gas-capes will be loaded on trucks by 0430 hrs.

C.S.M. will give necessary orders for "A" Echelon to report to T.O. at "B" Echelon by 0455 hrs.

"*B*" *Echelon* is in wood $\frac{1}{2}$ mile south of ESTAMINET ANGLAIS.

Weapons.—Platoon weapons.

Plus per Pl.—Three No. 73 Grenades.

Twelve No. 69 Grenades.

Three S.I.P. Grenades.

Equipment.—Respirator in wear. Haversacks, gas-capes on trucks; rubber shoes in wear.

Ammunition. ———

R.A.P.—Hospital 123456.

Intercommunication

Battalion H.Q.—Hospital 123456.

Coy. H.Q.—House No. 40; will move on 1st success signal via Street A. to House No. 15; on 2nd success signal to SCHOOL.

Signals.

Recognition Signal (to be hung on all captured houses).—Piece of material hanging from one window in each wall.

Success Signals.

Phase 1.—Green, White, Green, fired by O.C. No. 8 Pl.

Phase 2.—White, Green, White, fired by O.C. No. 7 Pl.

To Battalion.—Red, Red, Green, fired by O.C. No. 1 Coy.

Synchronize watches.

Any questions?

D.S. SOLUTION TO QUESTION 2

[QUESTION 2.—*As O.C. No. 8 Pl. give your orders for the first phase of the attack only.*]

Information

Enemy.—You have seen what opposition we have had so far. Nothing further is known of the enemy.

---

[7] The C.O. laid down that companies would be fed *by* 0430 hrs. It is up to the Company Commander to say *at* what time they will feed.

Own Tps.—No. 1 Coy. is to capture the SCHOOL in two phases. I am dealing only with the first phase here.

Coy. attacks two Pls. forward, one in reserve: No. 7 Pl. is on our right, with objectives Houses Nos. 34, 31, 21, 20 and 15 and route through back gardens between Streets B and A. They cross Start Line at zero.

No. 9 Pl. is in reserve behind us.

No. 2 Coy. is in reserve behind No. 1 Coy. Coy. has under command one Det. 3-inch M. which is in reserve, and one Sec. Carrier Pl., whose Bren guns are giving roof-covering fire, and cordoning Streets B and C.

"Now, I'll just repeat that. . . ."

Intention
No. 8 Pl. will capture and occupy Houses Nos. 8, 9, 10 and 11.

Method
Pl. will attack two Secs. forward, one reserve. No. 1 Sec. Right, No. 2 Left, No. 3 Reserve.

Objectives.
   No. 1 Sec.—Houses Nos. 9 and 10.
   No. 2 Sec.—House No. 8.
   No. 3 Sec.—House No. 11.

Route.
   No. 1 Sec.—Street A.
   No. 2 Sec.—Street G between Houses Nos. 18 and 19 and 14 and 8.

Pl. Boundaries.—Right excl. houses on right of Street A.
                 Left excl. Street C.

Fire Plan.
   No. 3 Sec. will give covering fire to Street A from House No. 34.
   2-inch M. will fire smoke on road junction Streets A and D from $Z-30$ sec. to $Z+3$ min. from area House No. 42.
   A/T Rifle with No. 3 Sec. covering fire.

Action on Capture of Objective.—Success signal will be fired by O.C. No. 8 Pl. On success signal covering fire numbers will close on their Secs. at Sec. objectives.

Phase 2 starts on success signal.

F.U.P.—House No. 43.

Start Line.—Street F.

Zero Hour.—0525 hrs.

Administration

Weapons.—All Pl. weapons will be taken.

Pl. Sergeant will draw from Coy. H.Q. by 0445 hrs.
- Four No. 36 Grenades.
- Three No. 73 Grenades.
- Twelve No. 69 Grenades.
- Three S.I.P. Grenades.

Distribution.—Nos. 1 and 2 Secs., two No. 36 Grenades each. Each Sec., one No. 73 and four No. 69 Grenades. S.I.P. Grenades at Pl. H.Q. in reserve.

Equipment.—Respirators and rubber shoes in wear. Boots and haversacks will be loaded on truck by 0430 hrs.

Ammunition. ———

R.A.P.—Hospital 123456.

Intercommunication

Company H.Q.

Till our Success Signal.—House No. 40.

On our Success Signal.—Move via Street A to House No. 15.

Pl. H.Q.—House No. 39, move up Street A to House No. 9.

Signals.

Recognition (to be hung on all captured houses).—Piece of material from one window of each side.

Success.—Green, White, Green, to be fired by O.C. No. 8 Pl.

Synchronize watches.

Any questions?

D.S. SOLUTION TO QUESTION 3

[QUESTION 3.—*What message would you expect to receive back?*]
[QUESTION 3A.—*Who would bring it?*]

Question 3.—Carried by observer on left side of Street A:

"Enemy in fourth house along on right-hand side of road. House is the only one set back from the road. Enemy in near front window, top floor.

"The enemy are in position to fire down Street E.

"I could see no sign of wire or barricades.

"Covering fire cannot be brought to bear from the end of the street.

"No. 1 Sec. is in Houses Nos. 19 and 21."[8]

Question 3a.—One of the observers on the same side as the dead scout. He would be observing the opposite side—that is, the side from which the fire came.

---

[8] While the details of the reader's solution will, of course, vary, the contents should, generally speaking, be the same.

D.S. SOLUTION TO QUESTION 4
[QUESTION 4.—*What action do you take?*]

1. Send the observer straight to No. 2 Sec. in Street G, to tell them that Street E is covered by enemy fire, and to tell them to occupy House No. 19 to give fire on the enemy. To tell them that No. 3 Sec. may assault down Street A.

2. Tell No. 3 Sec. to stand by to assault, but to wait for the word.

3. Tell the Pl. Sgt. to send the information back to Coy. H.Q.

4. Go to No. 7 Pl. H.Q. and find out if they require an assault from Street A. (They will probably by this time be in the process of attacking House No. 21 from the rear.)

D.S. SOLUTION TO QUESTION 5

[QUESTION 5.—*As N.C.O. i/c No. 7 Sec., No. 9 Pl., what orders have you given your Sec. at the Assembly Position?*]

Information

Enemy.—You have seen what opposition we have had so far. There is no further information.

Own Tps.—No. 8 Pl. is giving covering fire and smoke from Houses Nos. 8, 9, 10 and 11.

No. 7 Pl. is occupying House No. 4 and the houses beyond the School. Nos. 8 and 9 Secs. are following behind to enter after us.

Intention

No. 7 Sec. will force an entrance to the SCHOOL opposite House No. 4 and hold one room.

Method

Route.—Street A to House No. 8; across Street D.

Formation.—Single file on left-hand side of street.

Section will be divided into:

- *Nos. 3 and 4* (assault party with No. 73 grenade).—You two will will double across Street D from House No. 8 into House No. 4, up to an upstairs room, and drop No. 73 grenades against the ground at the school side of the passageway. You will then double down and enter behind the remainder of the Sec.

- *Nos. 5, 6 and 7.*—As soon as the explosion takes place, you will double, at 10-yards interval, to gap in the wall. No. 5, you will throw two No. 36 grenades into the room. I will then enter; you will follow, then Nos. 6 and 7.

- *Nos. 1 and 2.*—You will remain at House No. 8 until you have seen us enter the house. You will give local covering fire. As soon as we have entered you will follow us.

Action inside the House.—I will take up a position covering the door. No. 7, you will open the door. I will at once fire a burst through into the hall. You will throw a No. 69 and a No. 36 grenade into the hall.

No. 2, you will hang material out of the window. Nos. 8 and 9 Secs. will then clear the rest of the building.

In case of casualties you will take command of the Sec. in the order of your numbers.

No member of the Sec. will cross Street D until the recognition signal is hung by No. 7 from House No. 4.

Administration

Weapons.—Pl. weapons as usual.

    No. 3.—No. 73 Grenade.*

    No. 4.—No. 73 Grenade and two No. 36 Grenades.*

    No. 5.—Two No. 36 Grenades.*

    No. 7.—One No. 69 Grenade and one No. 36 Grenade.

    No. 2.—Take a piece of cotton with you to hang out of the window.

    Bayonets fixed.

Equipment.—As you are.

Ammunition ———

R.A.P.—Hospital 123456.

Intercommunication

Pl. H.Q.

Coy. H.Q.

Signals.

Synchronize watches.

Any questions?

D.S. SOLUTION TO QUESTION 6

[QUESTION 6.—*Describe how, as O.C. No. 2 Coy., 1 WARWICKS, you would set about clearing the area allotted to you.*]

1. Distribution of Platoons

    i. One Pl. cordoning sub-areas from SCHOOL end, by occupying:

        (*a*) House No. 3.

        (*b*) S.W. and S.E. rooms of SCHOOL.

    ii. One Pl. clearing houses.

    iii. One Pl. in reserve.

---

* Remainder of grenades assumed to have been expended.

2. Order of Clearing Sub-Areas

Area Searched.	Houses Occupied (*as opposed to cleared*).	Cordoning Houses.
Bounded by Streets D, A, F and B.	Nos. 20 and 34.	S.W. and S.E. rooms of SCHOOL.
Bounded by Streets D, C, F and A.	S.W. and S.E. rooms of SCHOOL. Nos. 20 and 34.	Nos. 20 and 3.

3. Houses Held at End of Operation
Nos. 20, 34 and 3, and S.E. and S.W. rooms of SCHOOL.

4. Search Houses on following System:
  i. Roofs and exits of each searched house covered with fire.
  ii. Entrance made through least exposed side of house.
  iii. Never less than two searchers together.
  iv. Always a reserve in hand.

Other Questions Instructors might ask are:
1. Who attends the order groups of the C.O., Coy. Comd., and Pl. Comd.?
2. From where would the orders be given?
3. At what time would they be?
4. What would be the first actions of subordinate comds. on leaving an order group?
5. What, in a given situation, would a Warning Order contain?
6. What would, say, No. 3 of such-and-such a Sec. be carrying?
7. Give the necessary steps, the timings and the location of a unit before a given attack.
8. A direct hit wrecks the building you had announced would be Coy H.Q. What do you do?
9. You are the Point Sec. Comd. You are held up by L.M.G. fire from such-and-such a place, and cannot advance down your route. What do you do?
10. You are Comd. of a Sec. ordered to assault a given house. Just as you are in the act of assaulting the house you see the enemy leaving it and going into another house. What do you do?
11. You are Sec. Comd. of a Sec. ordered to cross a street which an enemy L.M.G. is thought to be covering. Do you make your Sec. cross singly or in a bunch?
12. You are an L.M.G. No. 1 covering a street. If you err at all, do you think it is better to aim too high or too low?

APPENDIX I

PROTECTION TABLE

Reprinted from F.S.P.B. Pamphlet No. 4, 1939, page 19, Amendment 1 (December, 1940) by permission of the Controller of H.M. Stationery Office.

MATERIAL.	NORMAL FIELD DEFENCES. Safe thickness in inches at 100 yds. against (*a*) S.A.A. up to 7.92 mm. (bursts of 5 rounds L.A. fire or single shots A.P.); (*b*) bomb splinters	DEFENCE AGAINST A.TK. WEAPONS. Safe thickness in inches at 100 yds. against light A.Tk. weapons to 10 mm.	REMARKS.
(1)	(2)	(3)	(4)
1. Brickwork in lime mortar	13½	27	Good quality brick.
2. Brick rubble confined between 1 in. boards	12		
3. Chalk loose, as in new parapets	24		Consolidation decreases protection.
4. Clay loose as in new parapets	36		Much depends on type of clay. Good factor of safety is allowed here.
5. Coal (hard), confined between boards	13		Unsatisfactory owing to pulverizing effect of bullets.
6. Coal (kitchen), ditto	18		
7. Concrete, reinforced	6	15	
8. Earth or loam as in parapets	36	60	
9. Road metal, 1½ in. to 2 in. between 1 in. boards	9		
10. Sand between 1 in. boards	12		
11. Sand, loose	24	48	Protection given by coarse sea sand is considerably greater.
12. Sandbags filled with—			Thicknesses are the lowest multiples of sandbag dimensions to give the required protection.
Brick rubble	20		
Chalk	20		
Clay	30		
Earth	30	60	
Road metal	10		
Shingle up to 1 in.	20	30	
Sand	20	40	
13. Shingle or broken stones between 1 in. boards	9		

APPENDIX II

THE STRENGTHENING AND ADAPTATION OF A HOUSE FOR DEFENCE

GENERAL.

It is not within the scope of this book to describe more than the principles and general method of shoring, etc., as applied to the usual domestic type of house construction. Buildings with basements, buildings of steel frame construction, or other types of building, require some variation of treatment. The main idea will, however, remain identical in all cases. Shoring must start from a sound, firm and solid foundation, and support the ceiling over the room to be occupied, so that it is proof against any burden which may fall upon it.

1. GLOSSARY OF BUILDERS' TERMS USED IN THIS APPENDIX.

Shoring .. .. Supporting and thereby strengthening a ceiling with timbers.

Plate .. Timbers used horizontally in house construction for taking bearing of other timbers. Usually 4½" × 3" in section.

Dead Shore .. Vertical strut or support used in shoring.

Head tree (or top plate) Top horizontal timber used in shoring.

Folding wedges .. Wood wedges for driving between the foot of each dead shore and the bottom plate to tighten up and obtain equal loading on to the struts.

Sole plate .. Bottom horizontal plate under the feet of the dead shores to obtain level and sound bearing surface. In emergency these can be dispensed with providing that there is hard, sound concrete over the floor area below original wood flooring, or where the actual floor is of concrete, tile, or similar.

Iron dogs or cramps Used to secure the head of struts to top plates and the bottom to bottom plates, made of, say, ½" diameter round iron rod bent at right angles at both ends and pointed for driving. See Diagram.

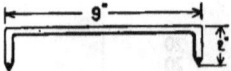

Braces .. Timbers used diagonally to tie struts together; can be floor boards or any similar and need not be thick timbers. Have the disadvantage of obstructing space in a room and are often dispensed with as not being really necessary if other timbers are properly fixed. The same result can be obtained without much obstruction by nailing boards to struts at just above head height and just above floor level.

Joists	Timbers carrying the floor boards and at right angles to them. Vary from 4" to 11" in depth.
Sleeper walls	Dwarf brick walls, 4½" thick, below ground floors and carrying floor joists on plates.
Chimney Breasts	Additional thickness of brickwork that surrounds a fireplace and allows of formation of flue. Normally projects into the room, but is occasionally flush inside and projecting outside where fireplaces are on external walls.
Party wall	Wall dividing rooms or adjoining houses. When dividing two houses it is normally not less than 9" thick.
Pinch bar	Stout iron lever tool for prising up flooring and timbers.

2. SHORING.

Stores.

Folding wedges	12 (2 per shore). Can be cut from timbers found, but would be best prepared beforehand and stored.
Iron dogs	36 (4 per shore).
3", 4" and 6" nails	2 lb. of each.
4 lb. hammer	1
1 lb. hammer	1
Handsaw	1
Pinch Bar	1
Pickaxe	1
Shovel (Navvy)	1

It is essential in shoring up to start from a firm foundation that cannot in itself collapse under an additional weight and that will not sag or sink. This means therefore that in an ordinary room with boarded floor the floor must be removed so as to allow the weight to bear on to a solid surface. Where the floor is itself of concrete, or tiles on concrete, or similar, this can be the foundation for the shoring work.

Normal type of wood floor in domestic house property is likely to consist of (*a*) floor boarding, (*b*) joists (usually of 4" × 2" section), (*c*) plates, and (*d*) sleeper walls.

To remove this type of flooring quickly, cut or prise up one or two boards with pinchbar or pickaxe or saw. Using the joist then exposed, prise up remainder of flooring, remove joists and plates and sleeper walls, all of which can normally be knocked apart with a heavy hammer. If joists are found to be "framed"—*i.e.*, jointed together—where running round the hearth, cut the joist at the front edge of the hearth in two places, when all will come up easily.

Usually the area under ground floors is covered with concrete to prevent damp affecting timbers, but in old property or property in rural districts, sleeper walls may have been built straight off the natural ground. Before any shoring is begun, this ground will require to be roughly levelled off so that the bottom plate of the shoring timbers has a level and sound surface underneath.

Hollow type floors, no matter what type of finished surface they have, can easily be distinguishable from solid floors by sounding.

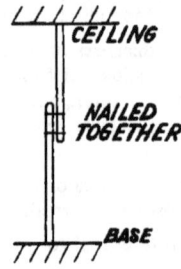

(i) Turn off gas at the meter and electricity at the main switch.

(ii) Remove the floor if not solid.

(iii) Make measuring rod. This need not be ruled in feet and inches, but should merely consist of two pieces of wood of sufficient length to measure the height of the room from ceiling to base. (See diagram, p. 51.)

(iv) Find timber and cut to required lengths. Wall plates and joists will probably be suitable, but should be examined for soundness, and nailed together to give the maximum possible sectional area for struts and top plate. Bottom plates can be 4½" × 3" wall plates or similar and do not require to be of great thickness; but the foundation under them must be level and firm.

Dead shores should be of minimum size of 6" × 4" (say, three 4" × 2"'s nailed together) and all cut to length taken from measuring rod, with allowance of 2½" for wedging.

Example.

Height of room from measuring rod	9 ft.
Thickness of top plate	4 in.
Thickness of bottom plate	3 in.
Allowance for folding wedges	2½ in.
Therefore the required length of dead shore is	8 ft. 2½ in.

(v) Ascertain direction of joists overhead by examining the floor of the room above or knocking down a small patch of ceiling. Joists will run the opposite way to the floor boards.

Top plate of shoring must run across (*i.e.*, at right angles to) the joists being supported.

(vi) *To erect Shoring.*—Place the bottom (sole) plates in position at each end of the room at a distance of not more than 12 in. from the end walls. Space other sole plates between, to give intervals not exceeding 5 ft.

Erect one end dead shore (again about 12 in. from the side wall), with wedges loosely inserted underneath. Place top plate on top of this dead shore and hold in position by loosely wedging up from bottom. Lift up the other end of the top plate into required position and repeat with another dead shore at the opposite end.

Erect intermediate dead shores (again at spacings not exceeding 5 ft.) and loosely wedge up.

All shores should be upright ("plumb").

Tighten up by driving the wedges in to grip firmly, but do not force to the extent of exerting lifting pressure on ceiling construction above, and see that the tightness is equal on all struts, so that some are not taking more load than others.

Secure all by driving in iron dogs at top and bottom, connecting top and bottom plates to the struts. If timber is available, cross-tie all struts, as described in "Braces" in earlier paragraph.

If any old corrugated iron or similar material is available, insert it between top plate and ceiling, making allowance for its thickness when cutting the struts. This would save possible injury from lumps of falling plaster and will give extra resistance to debris falling on the ceiling above.

3. ACCESS TRAP.

To obtain access from one room to a room below, it will be necessary to cut away the end of one joist (joists are normally spaced at about 12 in. apart). Select a position away from the fireplace on the first floor and then:

> Cut out a patch of flooring about 3 ft. square.
> Remove the ceiling from the underside and expose the joists.
> Securely nail stout timber across the joists, spanning three or four joists at least.
> Cut away the end of one selected joist, the end of which will then be held by the timber nailed across the others adjoining.

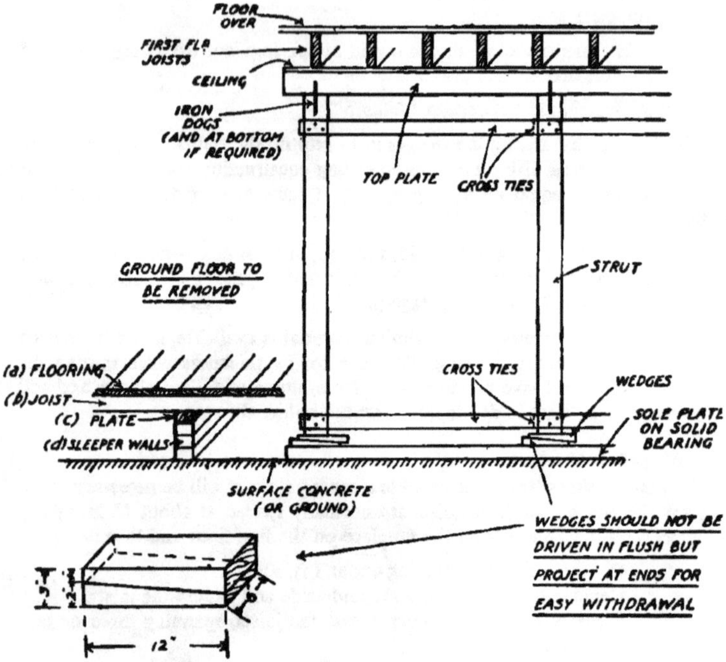

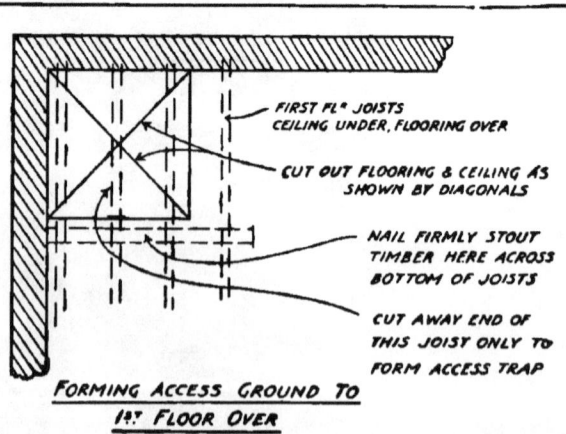

GENERAL IDEA OF SHORING 1ST FLOORS

FORMING ACCESS GROUND TO 1ST FLOOR OVER

APPENDIX III

SYLLABUS FOR A DEMONSTRATION

1. The wrong way to attack.
2. The right way to attack.
3. The preparation of a house for defence.

1. THE WRONG WAY TO ATTACK

No.	Point brought out.	Remarks	Unit.	Area.	Audience.
1	Protection wrong	Own troops in open without adequate sentries: a sudden attack finds them concentrated without protection, and leads to wild confusion.	Whole Pl.	CATTLEY'S	South end end of CATTLEY'S SQUARE
2	Protection right	Own troops under cover slightly dispersed by sections. O.Ps. and sentries covering every approach to hold an attack until main body can deploy.	Whole Pl.		
3	Cordon[1]	Cordon with fire, not men: to prevent enemy lateral movement, either of reinforcement, counter-attack or escape.	No. 3 Sec.	First floor window in KERRY'S	

However, after doing much damage, enemy retire and own troops advance.

| 4 | Scouts | 1. Look round corner and keep head there. Shot. No information except that enemy is somewhere in street. 2. Advance. Scout advances at double correctly, but comes to rest behind inadequate cover from view and is shot. | No. 1 Sec. | KERRY'S to WARRE SCHOOLS | WARRE SCHOOLS |
| 5 | Observers | Fail to come close to scout when turning corner and do not see what happens when scout is shot. | | | |

[1] Protection (right way) and Cordon (right way) are shown here to save the audience moving.

THE WRONG WAY TO ATTACK—*continued.*

No.	Point brought out.	Remarks.	Unit.	Area.	Audience.
6	Advance without covering fire	Enemy lies low and shoots scouts up from rear. Nobody to back them up.			
7	Bad covering fire	Man gets in obviously. Enemy either changes position so that he is ineffective, or blasts him while getting into position.			
8	Attack	1. Assault Sec. attacks up street before gaining houses on either side. Is shot up by enemy from opposite side who is covering doorway of attacked house. 2. Assault Sec. attacks through front instead of side which is defiladed from the street. 3. Remainder of Assault Sec. do not allow sufficient time for explosion, and do not run forward at 5-yard intervals. They are caught at the entrance, bunched together, and killed.			
However, some of them manage to penetrate into the front hall.					
9	Searching house	In front hall, N.C.O. stands and, as they come past, pairs them up and details them to search cellar, ground floor or upstairs. As he does this, grenade comes rolling down steps and they are all blown up.			
However, enemy are also killed, and own troops pass through street.					
10	Consolidation	They fail to leave anybody to cover the street, which is reoccupied by enemy.			

2. The Right Way to Attack

No.	Point brought out.	Remarks.	Unit.	Area.	Audience.
1	Covering fire	Platoon advances from Forming-up Position. Position a compromise between— 1. Being able to give covering fire till last possible moment; 2. Obtaining a long beaten zone; 3. Avoiding the obvious	No. 3 Sec.	HERBERT'S and KERRY'S	WARRE SCHOOLS
2	Scouts	Duty: to locate the enemy. One each side of road. Moving delicately, watching opposite side.	No. 1 Sec.	HERBERT'S to WARRE SCHOOLS	
3	Observers	Duty: to observe what happens to scouts, and watch for enemy movement. Always two on each side, keeping under cover as much as possible.			
4	Remainder of Point Sec.	Moving behind, spread out, ready to support, cover flank streets, search cul-de-sacs, alleyways, etc.			
5	Rear Sections	Under cover, close up, ready to support when required, but not exposed. Slightly dispersed. Observer watching with periscope.	No. 2 and 3 Secs.	CATTLEY'S	
6	Mopping up	Rear sections search houses, one section each side of street. Rapid, unthorough. Work in pairs. N.C.O. outside to control section and watch situation	No. 2 and 3 Secs.		
7	Consolidation	In clearing and holding an area only. Probably rifleman, possibly L.M.G. left to cover street, and prevent counter-action.			

THE RIGHT WAY TO ATTACK—*continued*.

No.	Point brought out.	Remarks.	Unit.	Area.	Audience.
8	Report	Pl. Comdr. at once reports back— 1. Street cleared. 2. Next action.			

Move farther along.

No.	Point brought out.	Remarks.	Unit.	Area.	Audience.
9	Action of Scouts when held up	Get under cover like lightning, and try to work into a position from which they can cover the assault.	No. 1. Sec.		
10	Action of Observers	One under cover to watch. The other back to report:— 1. Number and exact position of enemy. 2. Best approach. 3. Whether covering fire can get at them and where from. 4. Action of Point Sec.			
11	Orders	Given out over loud-speaker.			
12	Covering fire	1. Smoke. 2. S.A.A. 3. A/T rifle. 4. A.P. grenades. 5. S.I.P. Grenade. Back of house covered as well as front. Party attacking back of opposite houses first.	No. 1 and 3 secs.		
13	Assault	1. Explosives. 2. Heavy axe or crowbar. 3. Assault. 4. Entrance. Avoid the obvious. Have a man watching for grenades.	No. 2 Sec.		

(Consolidation, mopping up and further advance not shown.)

3. THE PREPARATION OF A HOUSE FOR DEFENCE

No.	Point brought out.	Remarks.	Audience
1	Stores	Wheelbarrow, Pickaxes, Saw, 4″ and 6″ nails, Iron dogs, Sandbags, Sand or rubble, Wire, barbed, Wire, plain } laid out on front pavement	Pavement opposite house
2	Choice of house	Choice of house is determined by— 1. Field of fire. 2. Defilade from enemy covering fire (back from road). 3. Strength of construction. (Not an isolated post).	
3	Glass knocked out	Remember to knock glass out of neighbouring houses at the same time.	
4	Inflammables and breakables cleared	Leave muslin curtains to cover windows.	Single file "one way street" through house, noticing points which are placarded in large letters.
5	Room shored up	See Appendix II	
6	Emplacement	1. Dimensions, inside, 6′ × 3′ 6″ × 5′ 2. Thickness, see Appendix I. 3. Field of fire. 4. Roof.	
7	Walls strengthened	Going on while shoring up and emplacement building is being done. Thickness, see Appendix I.	
8	Field of fire cleared	If this is obvious, duplicate with neighbouring houses.	
9	Loopholes	Widen main field of fire, cover side and rear of house. Duplicate.	
10	Barricades	Barricade all entrances and exits to house, including chimney, except places where a barricade would be visible from outside. Mine these or cover them with fire, or have many duplicated barricades.	
11	Mines and booby traps	If possible mine entrances not barricaded. Also lay booby traps inside house.	
12	Escape hole	Through wall into back of next door cupboard. In detached houses into lean-to shed, bush, etc.	

THE PREPARATION OF A HOUSE FOR DEFENCE—*continued.*

No.	Point brought out.	Remarks.	Audience.
13	Alternative positions	Put house close by into adequate state of defence.	
14	Dummy position	Wire through walls into house two or three away, where bolster hangs. Jerk when firing.	
15	Wire over windows	To stop grenades. Leave slit for dropping grenades out. Must be inconspicuous. Fine mesh.	
16	Fire precautions	Commandeer extinguishers from neighbouring houses.	
17	Covered retreat	If it is desired to retreat through back gardens, shift washing, bins, timber, etc., to make a covered line of retreat. Must be inconspicuous.	
18	Barbed wire	Place barbed wire where enemy will come upon it suddenly. Don't give away position of occupied house by siting of wire.	
19	Removal of clues	See no external signs of occupation left: 1. Trail of sand up to door. 2. Sandbags showing. 3. No other windows broken in street. 4. Direct view into defended room. 5. Loopholes unduplicated. 6. Absence of bushes, gate, etc., in front of house. 7. Garden heavily wired. 8. Wired window showing.	

Revised by F. J. O. Coddington, LL.D.(Sheff.), M.A. (Oxon), Stipendiary Magistrate, Bradford; Author, "The Young Officer's Guide to Military Law."

APPENDIX IV

SUPPRESSION OF CIVIL DISTURBANCES

HAVING considered street fighting in its ordinary military sense, it now remains to consider that special aspect of the subject which arises when military force is called upon to intervene in a civil disturbance for the purpose of reinstating law and order within a city.

It must be clearly understood that the subject is here discussed upon the supposition only of a disturbance within the British Isles. It is necessary at the outset to state this clearly because the subject is dealt with from a legal as well as a military point of view, and the laws and conditions under which military assistance may be invoked vary very considerably in different countries, and therefore to lay down a general rule upon the subject which could be applied in all countries alike would be impossible.

When a soldier is called upon to undertake operations against civilians he is at once confronted with three main considerations. These may be detailed as follows:

1. His relationship with the civil authorities.
2. The degree of assistance to be given, and the method whereby it is to be given.
3. The legal position of himself and his men in so doing.

RELATIONSHIP WITH CIVIL AUTHORITY

Whenever troops are called out by a civil authority to assist in quelling a riot or civil disturbance within the British Isles, the military officer to whom the request is made should whenever possible obtain a request in writing from the civil authority asking for his assistance.[1] If in response thereto a body of troops is dispatched, the commander of the party sent should be informed that it is his duty to report daily, and in writing, as to the progress of any operations which he may undertake, both to the War Office direct and to the officer commanding the station from which he is dispatched.[2] He should be informed that the civil authority of the area concerned is responsible for arranging for their adequate accommodation and food, and for meeting them upon arrival at the appointed rendezvous and conducting them to their quarters or to the point whereat their assistance is required.[3]

Throughout the period during which the troops remain at the seat of the disturbance, a magistrate and senior police officer should remain constantly in attendance, and the commander should from time to time consult with these

---

[1] See K.R., para. 1305 (a).
[2] See K.R., para. 1304.
[3] See K.R., paras. 1305 (d), 1306, 1307.

officials as to the best disposition of the troops. In the event of the troops being divided into more than one body, a magistrate and police officer should be requested to accompany each party.[4] The magistrate is not bound by law to go with the troops, if the danger is sufficient to deter a reasonably brave man, but he should be strongly urged to accompany the commander. It should be clearly understood by the military commander that the civil authorities are entitled to call upon him to help to suppress the riot, and he is bound to do so, but it is for the military authorities to decide on the strength and composition of the force [K.R., para. 1305 (e)] and for him to give all orders to the troops.[5] The civil authorities may request that certain action be taken, but further than that they may not go, and on no account may they be allowed to give orders to the troops. All ranks should therefore be strictly warned beforehand that they must not on any account take orders from the civil authorities.

The Act of Parliament which deals with the law appertaining to Civil Riots is an Act passed in the reign of King George I and now known as the Riot Act. The legal effect of the Act, so far as it affects the soldier, may be briefly stated as follows: If within one hour of the reading, or attempted reading, of a certain prescribed proclamation under the Act the (twelve or more) riotous persons assembled do not disperse they are guilty of a felony.[6] It should be accepted as a fixed rule that whenever possible this proclamation should be read by the Magistrate prior to troops being ordered to take action. In order that the attention of all persons present should be called to the fact that the proclamation is about to be read, its reading should be prefaced by "The Alarm" loudly blown on the bugle or trumpet [K.R., para. 1309 (b)]. It is important to note, however, that the reading of the proclamation (still less the waiting for an hour afterwards) is not essential if there exist such exceptional circumstances as to justify immediate action on the part of the military commander.[6] It must be understood, however, that a commander who acts without the previous reading of the proclamation must be prepared to show that his action was absolutely necessary in order to prevent the occurrence of an outrage of a felonious[6] and violent nature.

METHOD AND DEGREE OF SUPPORTING THE CIVIL AUTHORITIES

It should always be the object of a military commander who is engaged in the invidious task of assisting the civil police to restore law and order, to endeavour, in the first instance, to use his troops rather as a rallying-point for the police than to initiate an offensive against the rioters. The reason for this is clear, viz., that the object in view, and the means whereby it is sought to achieve it, are entirely foreign to the training and tradition of a soldier. In a civil disturbance the object in view is merely to disperse the rioters and restore order with the least possible amount of damage, and with the infliction of the fewest possible number of casualties. Such an object is not compatible with the normal mentality of the soldier, and it involves the introduction of a degree of gentleness into the ensuing operation with which the military creeds (such as

---

[4] See K.R., paras. 1307, 1308.
[5] See K.R., para. 1314.
[6] See *Infra*, Legal Position of Officer Opening fire, p. 62.

"the spirit of the bayonet," etc.) are not likely to be harmonious. It is essential that so long as the troops are not actually required, a commander should, as far as possible, keep them in places which are not exposed to the jibes and missiles of the rioters. It is of the utmost importance that the troops should retain their coolness and good temper, and every effort must be made to avoid the occurrence of untoward incidents, whereby they might become unduly angered and difficult to keep in hand when the time arrives for them to take a more active part in the proceedings. It must be borne in mind, too, that the mere sight of the military (even at a fair distance) with their arms and equipment will, as a rule, have a considerable deterrent effect on a crowd of unorganized civilians. If this alone does not suffice, however, and it becomes necessary to support the police more strongly, the military commander will best achieve his object by drawing a cordon of troops across the street or square in rear of the police, who, if overpowered and obliged to retire, can then fall back upon the troops. The troops will at once open their files and allow the police to pass through their ranks. In this manner a cordon of troops will be interposed between the police and the rioters, allowing the former to rally and re-form without hindrance.

Cases occur, however, when other measures fail, and it becomes necessary for the troops to intervene in a more definite fashion and for fire to be opened.[*] It is of the utmost importance in such an event that the men should be told to direct their fire upon the foremost ranks of the rioters, in order that those leading the disturbance shall suffer most. It should be clearly pointed out to them that to fire over the heads of the crowd will invariably involve the deaths of innocent people who are probably a considerable distance away from the actual scene of the trouble, and who are either not participating therein at all or only to a very mild extent, while those who are guilty of leading and encouraging the riot are left unhurt, free to continue the fomentation of the trouble (K.R., para. 1320).

THE ORDER TO FIRE

The question of "the order to fire" is here separately considered owing to the many disputes and controversies which have raged round it in the past. Many commissions, judicial inquiries and inquests, when concerned with the investigation of incidents which have occurred in the course of conflicts between rioters and troops, have found that their greatest difficulty has been to determine by whom and under what circumstances "the order to fire" was given. It is essential, therefore, that all troops before being sent out to take part in the suppression of civil disturbances should be clearly told beforehand not only the name of the officer from alone they will take orders to open fire, but also the name of the officer who will exercise that power in the event of the first-named officer being injured or absent. It should be impressed upon the troops that they will not open fire except by command of the officer or officers so named.

The regulations laid down in King's Regulations, paras. 1315 to 1318, which deal with the telling-off beforehand of sections for the purpose of opening fire should necessity arise, and which provide for the control of fire when actually opened, must be strictly observed. When a request is made by a magistrate for fire to be opened, a commander should, where possible, insist for its own protection upon it being made (or, at any rate, immediately afterwards confirmed) in writing.' For, although such a request will not of itself suffice entirely to absolve an officer from the consequences of his act in opening fire should his conduct in doing so be the subject of subsequent investigation, yet it will certainly be taken as strong evidence in his favour, and will go a long way towards justifying the action he took.

LEGAL POSITION OF OFFICER OPENING FIRE

It is here proposed very briefly to consider the legal position of officers and other ranks engaged in the suppression of civil disturbance. It may safely be laid down that troops should never fire on civilians unless they are *rioting*. The legal definition of riot is that it is "a tumultuous disturbance of the peace by three or more persons who assemble together, without lawful authority, with an intent mutually to assist one another against any who shall oppose them in some enterprise (of a private nature), and who afterwards *actually begin* or execute the same *in a violent* or turbulent *manner* to the terror of the people."

By an "enterprise" is meant some purpose to destroy some building, or to kill or seriously injure some person, or the like.

"Of a private nature" is not important for this purpose. If the enterprise is of a public nature, the "riot" turns into an insurrection or "levying war against the King," which justifies the use of force to repress it even more obviously than a riot does.

"To the terror of the people" means so as to alarm a reasonably normal man.

The point at which extreme force (such as firing) may become justifiable is:

(a) Before the proclamation (or attempted proclamation), or before the expiration of the statutory hour after: if the rioters proceed to begin felonious violence. The violent felonies most likely to occur are: arson (burning) of a building; demolition or mass looting of a house, factory, town hall, law court, prison, etc., murder or serious wounding, etc., of one or more persons.'

(b) *After the statutory hour*: those remaining become guilty of felony, and all necessary violence may be used to disperse them.

(c) At any time: if the rioters attack the troops, in self-defence.

Even in these cases it must be remembered that in the eyes of the law the military man is only a citizen armed, and is personally responsible for his actions. If he has time and opportunity he must use his own discretion. Only if he has not is he entitled to accept the statement of the facts (*e.g.*, as to the degree of danger) from the civil authorities. He must fire if, but only if, he cannot otherwise stop the violence which is being (or is manifestly about to

---

' See K.R., para. 1311.
' For full list of *felonies* see table at end of Chapter VII, Manual of Military Law.

be) committed. The essential fact to be realized by all ranks is that troops are only allowed to use such a degree of force as, and no more than, is sufficient to quell or prevent the continuance of the riot or disturbance. On no account may a greater degree of force than is absolutely necessary be used. Any officer or man who uses force recklessly, or who prolongs its use unnecessarily, is indictable before a criminal court for manslaughter should casualties result. This is true even though he may be obeying the order of a superior officer in so doing, for in such a case it would be under civil law that prosecution would be taken, and when a military command is at variance with civil law, civil law will prevail. In practice, however, it would doubtless be considered in mitigation of sentence (though not accepted as an actual excuse) should it be proved that the action on the part of a soldier, which subsequently became the subject of criminal proceedings, was done by him in obedience to the orders of a superior.

Conduct of Troops

It should be impressed upon all troops taking part in operations which bring them into contact with civilians that their conduct must be such as will bear the powerful searchlight of publicity which will almost inevitably be brought to bear upon it after the conclusion of the trouble. They must have the fact impressed upon them that they have a serious and difficult duty to perform, and one which is not only distasteful to all soldiers, but is in many respects contrary to both their training and tradition. They must nevertheless perform it coolly, and, above all, with tact and forbearance, preserving at all times their steadiness and dignity as British troops.

Note

Acts which constitute a riot when done by civilians of course equally constitute a riot when done by soldiers, even if they take place in a military camp. Such an event is an example of a case where an officer must act, even to ordering troops to fire, without request from the civil authorities.[*]

---

[*] See K.R., para. 1322.

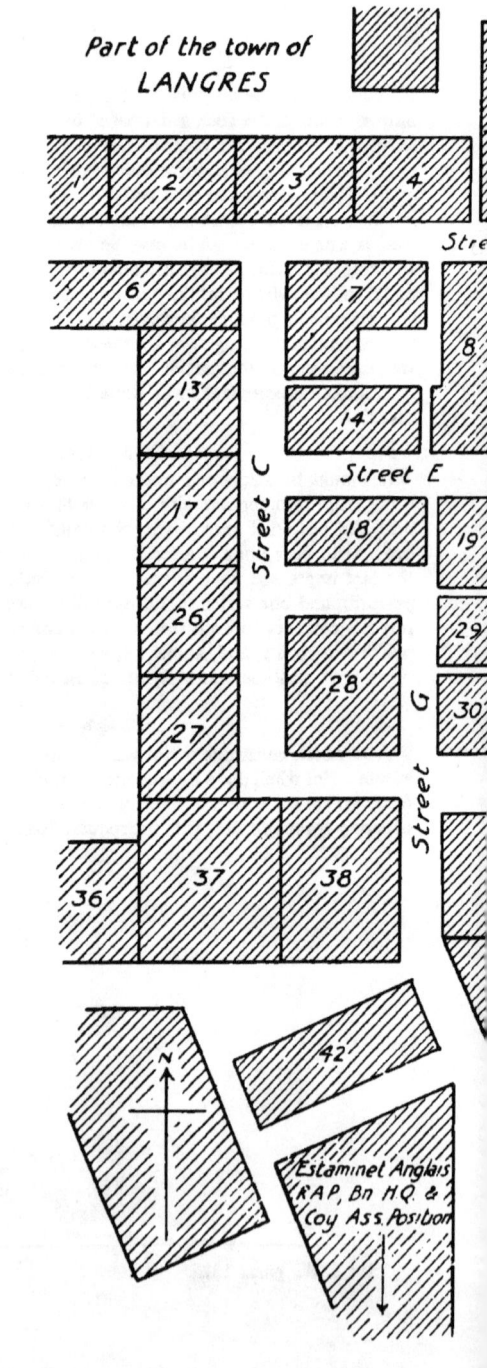

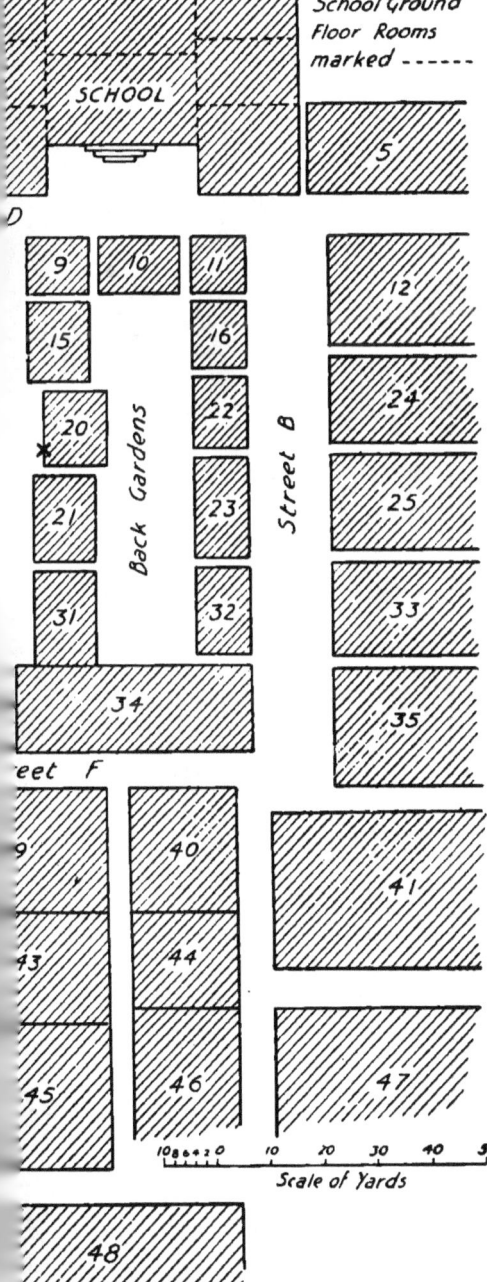

www.ingramcontent.com/pod-product-compliance
Lightning Source LLC
Chambersburg PA
CBHW060215050426
42446CB00013B/3080